THE HISTORY OF CLOCKS AND WATCHES

Clock tower of St. Mark's Square, Venice, showing the twenty-four hour dial and the jacks for striking the hours (see page 21)

THE HISTORY OF CLOCKS AND WATCHES

Kenneth F. Welch

Drake Publishers Inc : New York

To
Joan and Hilary Anne

ISBN 87749-242-5

Published in 1972 by
Drake Publishers Inc
381 Park Avenue South
New York, N.Y. 10016

Printed in Great Britain

Contents

List of Illustrations

1
The Sun and Stars

Primitive man was a hunter and his life was regulated by night and day. Then he learned how to grow crops, and he recognised that at certain times crops should be sown and at others they should be harvested. These were the seasons and he studied the stars, the moon and the sun to learn more about what we call Time. However, as life became more complex there arose the need for a greater understanding of time. During the daytime the first measurer was probably the shadow caused by the rays of the sun falling upon one or more fixed marks.

At night a bright star was chosen as a marker, and it was found that the time taken for the earth to make one complete revolution on its axis is almost constant; whereas the interval of time between two successive transits of the sun is irregular. This is because the sun's motion is not quite uniform. Thus, by comparing time measured across their markers by the stars (which is called sidereal time) and the time measured by the sun (which is called solar time), the early astronomers discovered that the two times differed.

Astronomers chose the sidereal day as the unit of time because it is constant; but everyday life is governed by the sun and not by the stars. However, if time is based on the sun it cannot be accurate because of the latter's irregular motion. So an imaginary sun was invented called 'mean sun'. The difference between mean solar time (indicated by the mean sun) and true solar time (indicated by the true sun) is called the 'Equation of Time'. In simple language it is man's way of balancing his accurate imaginary sun with the true sun. It is only on four days in the year that solar time and mean time agree – about 15 April, 15 June, 31 August and 24 December.

The solar year consists of 365 days 5 hours 48 minutes 48 seconds, and we reckon our year as 365 days, so we are nearly six hours short of the solar year. This we try to correct by adding an extra day every four years (leap year), although this correction is still too great, so the leap year is omitted every hundred years except in years which are multiples of 400. Because of these numerous irregularities it was agreed that the standard should be the mean of the solar day, and it is called Mean Time.

Long before these calculations were made rough time-keepers had been made for local use.

CHINA

About 3000 BC Fo-hi, the Father of China and its first emperor, studied the stars and gave his people a simple clock (see page 3) by which they could divide day and night into fairly equal parts. A metal rod covered with tar and sawdust stood on legs a few inches above a large metal plate. Two copper balls were suspended by thread over the rod. One end of the rod would be ignited and the moving flame would burn through the thread, thus causing the first ball to fall with a clatter upon the plate beneath. After a short interval of time the second ball would fall. This interval between the falling of the two balls could, of course, be varied by moving the threads either closer together or further apart.

This clock was not very accurate for the amount of sawdust and tar on the rod, the thickness of the thread and, of course, the draught would vary. But it was good enough to satisfy the Chinese, who did not need to be as accurate over their time as we are today.

ANCIENT ORIENTAL CALENDAR CLOCK

Later, the Chinese measured the length of the seasons, months, days and hours by the movement of the heavenly bodies. The path taken by the moon was divided into twenty-eight mansions or halting places and the sun was thought of as the hand of a huge clock. Its path was shown by twelve discs, each of which was given the name of an animal. The night discs, or yin discs, were black; the day discs, or yang

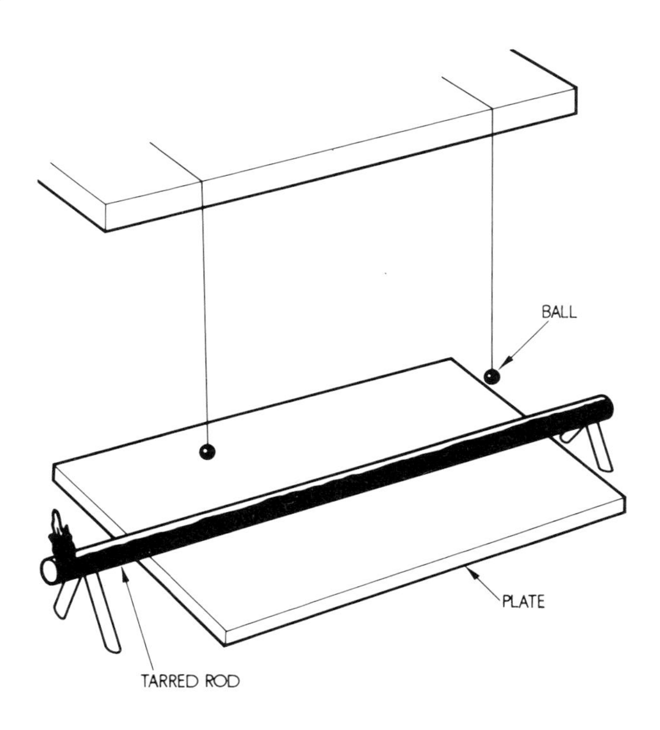

Clock of Fo-hi, first Chinese emperor, about 3000 BC

discs, were white. Grey discs showed the sunrise and sunset. Midnight was called *rat* or the dark hour. *Horse* was noon; *before horse* was the forenoon or morning, and the symbols dragon and snake were used; *after horse*, the afternoon, had the symbols sheep and monkey. Each time the sun had reached evening, or *dog*, the moon was in a new mansion of heaven.

The four great quadrants or quarters had colourful names: Azure Dragon, Black Warrior, White Tiger and Vermilion Bird. The indicator showing the season was named Peh Tao.

(We call it Ursa Major – the Great Bear – or the Plough, and it is used to find the North Star, which is in a straight line with the Plough and two other stars.) It was springtime when the handle of the Plough pointed east at nightfall, winter when it pointed north, summer when it pointed south and autumn when it pointed west.

BABYLON AND EGYPT

Babylonian priests told Alexander the Great that their records went back four hundred thousand years, and that more than two thousand years before the birth of Christ their astronomers had fixed and named the chief constellations that marked the annual path of the sun. With their knowledge they were able to verify, to within a few years, the solar system position such as would be indicated on a sundial.

About 500 BC a famous Babylonian, Nabu-rimanni, called Naburianus by the Greeks, made use of a sign for zero. He used it in his astronomical tables for the calculation of the new moon and eclipses. He was able to measure the time taken by the earth to rotate on its axis (this we call the day) and the time taken by the earth to revolve round the sun (this we call the year). From this he was able to calculate the length of the year in terms of days and parts of days. He arrived at the astonishingly accurate figure of 365 days 6 hours 15 minutes and 4 seconds. Today we estimate the year as being 26 minutes 55 seconds longer than this.

The Babylonians were greatly helped in their study of the heavens by their climate. In the clear skies they observed the movements of the planets and recorded results of their observations. Records of eclipses of the moon and sun and stars were so carefully kept that part of the ancient chronology has now been fixed without fear of mistake by just such occurrences. Even modern astronomers have not been able to accumulate a series of astronomical observations as long as those of the Babylonians. The longest known series of modern observations, at Greenwich, began only in 1750. The Babylonians had crude records for hundreds of years before their official series began. Those early astronomers probably looked at the stars because they wanted to read their future in them. Doubtless they soon realised from their observations that their studies would help in calendar making and time adjustment.

The accuracy of their calculations is due to their advanced mathematical knowledge. By 2000 BC they had already formulated the fundamental laws of mathematics. Although they knew all about the decimal system (based on the primitive method of using the fingers for counting), for complicated work they evolved the sexagesimal system. In this method the unit was sixty, not ten. It was successful and was gradually adopted in many parts of the world. It is still used today. We divide the circle into 360 parts, called degrees; the hour is divided into 60 minutes, and there are 60 seconds in a minute.

Diodorus of Sicily gives us some interesting facts about Egyptian time measurement in his writings, which date from about 50 BC:

> *The priests of the Egyptians, reckoning the time from the reign of Helos to the crossing of Alexander into Asia, say that it was, in round numbers, 23,000 years. And as their legends say, the most ancient of the gods ruled more than 12,000 years, and the late ones not less than 300 years. But since the great number of years surpasses belief, some men would maintain that in early times, before the movement of the sun had yet been recognised, it was customary to reckon the year by a lunar cycle. Consequently, since the year consisted of 30 days, it was not impossible that some men lived 1200* years, *for in our time, when one year consists of twelve months, not a few men live over one hundred years.*

The Egyptians had a calendar of red sandstone (page 6, number 1). There were three rows of holes with ten holes in a row, and the days of the month were recorded by moving a wooden peg from hole to hole. In the Egyptian year there were three seasons each of four months. The clocks illustrated here are examples of time-keepers which could be used

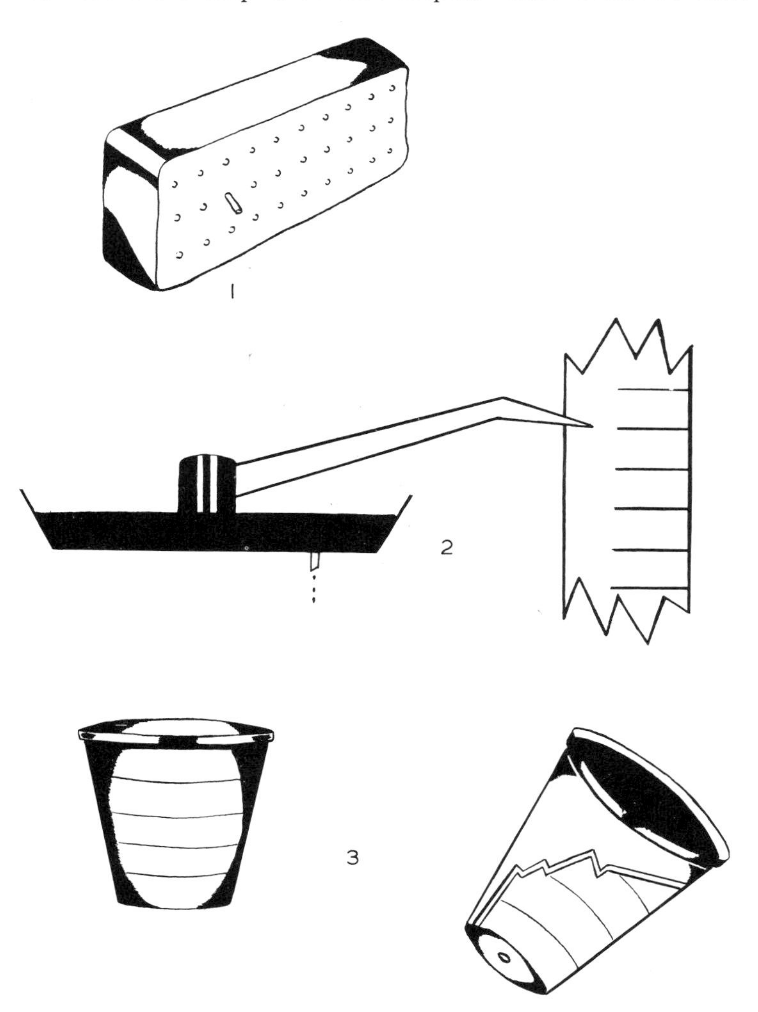

(1) Egyptian sandstone calendar
(2) Egyptian clock with pointer
(3) Egyptian water clock

day or night. Note the cylinder floating in a bowl of water (page 6, number 2). As the water drained away so the cylinder sank and the pointer recorded the time on a numbered board. The water clocks (page 6, number 3) were crude but effective. Inside the vessel, which was like a modern flowerpot, a number of lines or notches were scored. As the water emptied through a hole in the base of the pot these lines became visible and so denoted a passage of time. Owing to lime deposits the hole would gradually become smaller and this would cause the clock to be inaccurate. The sinking bowl was used for small divisions of time. In the base of the bowl was a small hole which let in a steady trickle of water, and the time taken to fill the bowl and sink it defined a required period of time.

The Egyptian shadow clock was of simple construction. The sun threw a shadow of a crossbar on to a longer horizontal bar, which was graduated to show intervals of time. Another important clock was the Clepsydra. Types of clepsydra were used 6,000 years ago in China, and in India in 300 BC. The Indians called theirs the jala-yantra. In Alexandria of Egypt, Ctesibius made a clepsydra (page 8) which is said to have been accurate and was self-adjusting. Water was poured into the small funnel and thence into a reservoir. In this cylinder was a float which held a thin pillar. At its top stood a figure of a man pointing to marks on a column. When the tube T was full it overflowed and so the float fell rapidly to the bottom of the reservoir. So far equal length hours would have been measured, but the Egyptians required a better method to show their hours which were of different lengths. Ctesibius was able to do this by drawing the hour divisions on the column on a slant so as to vary the distances between the lines of different sides of the column. When the water overflowed down T it entered a drum which revolved as each of its compartments became filled. The same principle is used in the modern water mill. In the illustration it will be seen that this drum was connected to the hour column by means of wheels and a tall spindle. In this way the column revolved and the man's pointer went up and down.

Although Ctesibius produced an efficient instrument for measuring time, a considerable amount of work had been done on time measurement in other parts of the world.

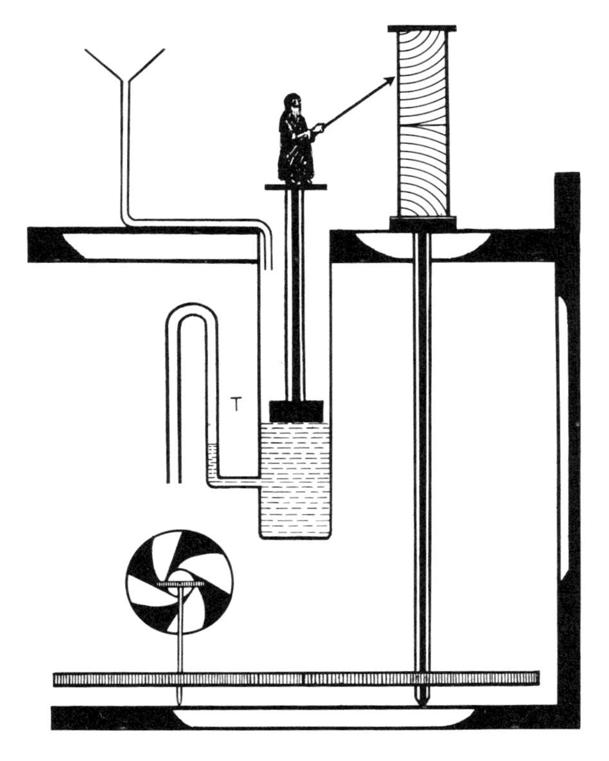

The clepsydra of Ctesibius, Alexandria

AMERICA, THE PACIFIC AND INDIA

The Incas of Peru, the Aztecs of Mexico, the South Sea Islanders and the ancient inhabitants of Britain all worked out astronomical methods for measuring time. Some astronomers plotted the course of the sun, others mapped the paths of selected groups of stars.

To the Hindu, time is a revolving circle. Time has no

beginning, no end. Hindus believe that all things in our universe (including the gods, or God) are bound together within a constantly repeated cycle of time. Time, they say, is divided into extremely long cycles of creation corresponding to the ten incarnations of one of their gods, Vishnu.

From all the data available it has been shown that the following extract from the *Book of Amos* in the Old Testament tells of an eclipse which was witnessed in the Middle East on the 15 June, 763 BC.

'. . . and it shall come to pass in that day, saith the Lord, that I will cause the sun to go down at noon, and I will darken the earth in a clear sky.'

GREECE

There was no *national* time in Greece and each city state had its own. This must have been most inconvenient to the surrounding peoples. Sometimes they repeated a sixth month and sometimes a twelfth; the year began on different dates in different places and the lengths of the months were decided by the local authorities. By the end of the fifth century BC the Greeks came to the conclusion that their calendar had not fulfilled its purposes. From their epics we learn that the Greeks had rather confused ideas about their own history. One simple method of recording the passage of time was the Olympiad. Every four years warlike pursuits were forgotten and the Greeks held the Olympic Games. The Olympiad dates from 776 BC and this system was used by the Sicilian historian Timaeus, a friend of Plato, to record the events of the different cities.

The astronomers could do better than Timaeus and from 747 BC a continuous record of solar and lunar eclipses was kept in Mesopotamia. Later taken over by Hipparchus of Rhodes in the second century BC, the compilation of the record went on until it was continued by Claudius Ptolemy of Alexandria (about AD 150), one of the greatest astronomers of ancient times. From the information it provides, it has been possible to date accurately events recorded in

Mesopotamia and elsewhere from the middle of the eighth century onwards.

The Greeks also designed and made a time-measuring

Arabian astrolabe

instrument called an astrolabe. It consisted of a circular disc suspended from a point on its circumference. A pointer with sights turned upon the centre of the circle as an axis and was directed towards the sun. The pointer could be adjusted to show the time of the day on the rim of the disc. The success of this instrument owes much to early astronomical research and to the fine craftsmanship of the Middle East around AD 1000. Later on the astrolabe was regularly used in Western Europe and towards the end of the fourteenth century Geoffrey Chaucer wrote a treatise on the instrument. The astrolabe gave latitude and time of day and was thus valuable to navigators until it was replaced in the 1700s by the quadrant. The astrolabe was small and portable and might be described as the first watch.

Not until the seventh century BC did the Greeks have a national calendar. It was a soli-lunar (sun-moon) calendar which needed intercalation. The Greeks worked on an eight-year group to which three extra months were added. This was inaccurate so in 432 BC Meton of Athens proposed the use of the Babylonian cycle in which a thirteenth lunar month was added seven times in a cycle of nineteen years. However, the system was not actually adopted.

THE ROMANS

Early in their history the Romans based their year upon the lunar month and their earliest calendars had ten months in a year and not twelve. However, later on, Numa Pompilius completed a calendar of twelve months, the year beginning in the spring (Vernal Equinox). Only six of these months were given names, the others being numbered. This did not suit the logical Romans, for six of the months would have an even number of days in them, and that was considered unlucky. Therefore, instead of alternating with twenty-nine and thirty days in a month they alternated with twenty-nine and thirty-one days, or sixty days to two months.

This, of course, gave a year of 360 days and fell short of the solar year by $5\frac{1}{4}$ days. But the Romans were more interested in the lunar year of 354 days, so they modified their year to 355 days. This they did by reducing the days in February from thirty-one to twenty-eight and only having twenty-nine days in December instead of thirty-one. The beginning of the year was changed from 1 March to 1 January.

Julius Caesar and Augustus both produced new calendars. The former abandoned the lunar month, took the solar year to be $365\frac{1}{4}$ days and called the civil year 365 days. Each fourth year he had an extra day added to it to account for the quarter day mentioned above. 23 February, repeated as the *leap day*, was regarded as the sixth day before 1 March. Augustus merely modified Caesar's calendar, as can be seen from the table.

Month	*Julian*	*Augustan*
January	31	31
February	29 or 30	28 or 29
March	31	31
April	30	30
May	31	31
June	30	30
July	31	31
August	30	31
September	31	30
October	30	31
November	31	30
December	30	31

As well as the astrolabe the Romans used the clepsydra, which had been brought from Greece in 157 BC by Scipio Nasica.

The Julian calendar was used for more than 1500 years in the Christian part of the world. The Gregorian calendar was designed to correct errors in the Julian calendar. Roman Catholic countries accepted it at once but England did not do so until 1752.

EARLY BRITAIN

While the Egyptians and the Greeks were busy with the sinking bowl, shadow clock and clepsydra, the less advanced Britons were working out part of the difficult task of making a sundial. A very early instrument for the measurement of time has been dug up at Camp in Staffordshire. It consists of four stones placed at the four main compass points. A leaning stone crosses diagonally the space formed by the four stones. This huge sundial is so built that it can mark the turning points of the yearly path of the sun. Shadows falling on prominent points or edges of the monument register the hours. Perhaps this instrument was used to indicate

Stonehenge as it might have appeared about 1400 BC, by Alan Sorrell

certain times of the year for some religious festival or for agricultural purposes.

Further proof that the ancient people in Britain knew something of the yearly passage of the sun and were conscious of a pattern in the seasons is found in Stonehenge in Wiltshire. This huge and mysterious place, where even quite recently new discoveries have been made, was so built that at the summer solstice the rising sun, resting its rays on the Heel Stone, could shine through the stone circles on to a flat stone in the centre. This stone is often called the Altar Stone. That Stonehenge and Woodhenge nearby were temples of some kind appears fairly certain. However, no one can say for sure whether the significance of these places was greater as temples for religious purposes or as instruments for measuring time other than at mid-summer.

SUNDIALS

Sundials, which undoubtedly existed as early as 2000 BC, probably took the form then of a stick standing upright, or an obelisk. The invention of a proper sundial is only possible when such an accumulation of information as the Babylonians had is available.

The earliest record of sundials is found in the Bible, 2 Kings XX – about 741 BC. When Hezekiah asked Isaiah what sign there would be that the Lord would heal him, the answer was:

> *'Shall the shadow go forward ten degrees or go back ten degrees?' Hezekiah answered, 'It is a light thing for the shadow to go down ten degrees; nay, but let the shadow return backward ten degrees.' And Isaiah the prophet cried unto the Lord; and he brought the shadow ten degrees backward, by which it had gone down in the dial of Ahaz.*

A Chaldean astronomer about the year 300 BC invented a sundial known as a hemicycle. A block of stone was hollowed out, and from the bottom of this hollow there rose a style or gnomon, which would throw a shadow across the curved interior. Each curve cut in the stone was divided into twelve pieces, which represented the hours between sunrise and sunset, and the hour marks were joined to give a series of hour lines.

In Saxon and Norman England sundials were very popular especially in the southern part of the country. Of all the hundreds which can still be seen, usually on the south walls of the churches, it is unlikely that there are two alike. They vary in size, shape, detail and position. There are wheel dials, half wheel and quarter wheel dials. The pointer (gnomon or style) sticks out from the dial and provided it is long and fat enough to give a good shadow it may be square or round. At Kirkdale Church in Yorkshire there is an interesting Saxon sundial dating from about AD 1064. The Saxons divided their sundials up into eight 'tides', of which only four could be recorded in daylight. The divisions were equal and therefore rather inaccurate. The Saxon day was as follows:

4.30 am	to	7.30 am	Morgan
7.30 am	to	10.30 am	Dael-mael
10.30 am	to	1.30 pm	Mid-daeg
1.30 pm	to	4.30 pm	Ofanverthrdagr
4.30 pm	to	7.30 pm	Mid-aften
7.30 pm	to	10.30 pm	Ondverthnott
10.30 pm	to	1.30 am	Mid-niht
1.30 am	to	4.30 am	Ofanverthnott

Here is a translation of the Saxon words on the sundial at Kirkdale Church.

Above the dial – *This is the day's sun marker.*

Below the dial – *Hawarth made me and Brand priests.*

From the inscription on either side of the dial we learn something of the church itself.

Orm, son of Gamal, bought St Gregory's Church when it was all broken and fallen down, and he caused it to be made from the ground, to Christ and St Gregory in Edward's days, the King, and in Tosti's days, the Earl.

Tosti was King Harold's younger brother.

French column sundial surmounted by the figure of a dragon, 17th century

WICK AND LAMP CLOCKS

The wick clock was another early method of measuring time. Knots were tied at equal distances apart in a length of wick, which was set alight and allowed to smoulder. Many years later King Alfred improved upon this idea. He used twelve-inch long wax candles marked off in inches. One candle lasted about four hours, but the King soon realised that the speed of burning varied with the amount of draught. So he invented a lantern to hold the candles, and instead of glass thin horn was used. As late as the eighteenth century wick and lamp clocks were still in use.

Until around AD 960 there were no mechanical clocks, only sandglasses (similar to the hourglass egg timer), sundials, clepsydrae, the candle and the graduated oil lamps. It is known, however, that by 1360 at least one mechanical clock existed. No one knows who made it and it is doubtful whether any one person was responsible. Probably it was a natural development.

In the ninth century a water clock of inlaid gold was given to Charlemagne by the King of Persia. The dial had twelve doors which represented the hours. Each door opened at the appropriate hour and the corresponding number of little balls fell out and dropped, one by one at regular intervals, upon a brass drum. When it was twelve o'clock twelve miniature horsemen issued forth and shut all the doors.

Between the tenth and fourteenth centuries, the clepsydra became more intricate. The hours were struck, tiny lions roared and small figures moved about. All these additions required small pieces of mechanism of some kind, and from their use men learned how to fashion a clock which could be wound up. If there was no water available then something else had to be provided to act as a driving power. Weights were used, but a difficulty arose because there was no way of controlling the speed at which they fell. The following experiment illustrates the problem, and many attempts at solving it must have been made in the fourteenth century.

Round a cotton reel wind a length of string, and on the

loose end tie a small weight. Through the hole in the reel slide a stiff piece of wire to allow the reel to revolve freely. When the weight is freed it unwinds the string rapidly, too rapidly in fact to be of any use. The problem then is to find some practicable method of slowing down the speed of the falling weight.

It can be done by fixing two little fan blades to the cotton reduce it enough, however, would require rather larger blades and would thus make the mechanism too cumbersome.

THE FOLIOT

An unknown inventor found the answer by making what is called a *foliot*. It consists of a crown wheel, cross-bar and two weights, and an escapement. The object of the foliot is to check, in an orderly manner, the running of a train of geared wheels. The revolving of a wheel with triangular teeth (called a *crown wheel*), which was the last and quickest-moving wheel of the train driven by the weight, was checked by two pallets (like the fan blades on the cotton reel) fixed to a vertical rod called a *verge*. These pallets alternately caught up first one tooth of the crown wheel and let that escape, and then another tooth on the opposite side of the wheel and let that escape. The teeth of the crown wheel thrust the pallets first one way and then the other. This caused the verge to oscillate backwards and forwards, thereby releasing the train of wheels in a step-by-step movement. To prevent this thrusting force running wild and the weight taking complete control of the mechanism, a cross-bar with two small adjustable weights hanging from it was fixed to the upper end of the verge. Thus the crown wheel now had to oscillate the verge and the weighted bar. If the weights were moved towards the ends of the bar the movement slowed down and so the clock went more slowly; when the weights were moved nearer the centre of the bar it went faster. The cross-bar and the two weights are called the *foliot*. In the diagram only the crown wheel is shown. The crown wheel, or escape wheel, and the two pallets are termed the *escapement*.

This system of controlling a falling weight was used by a

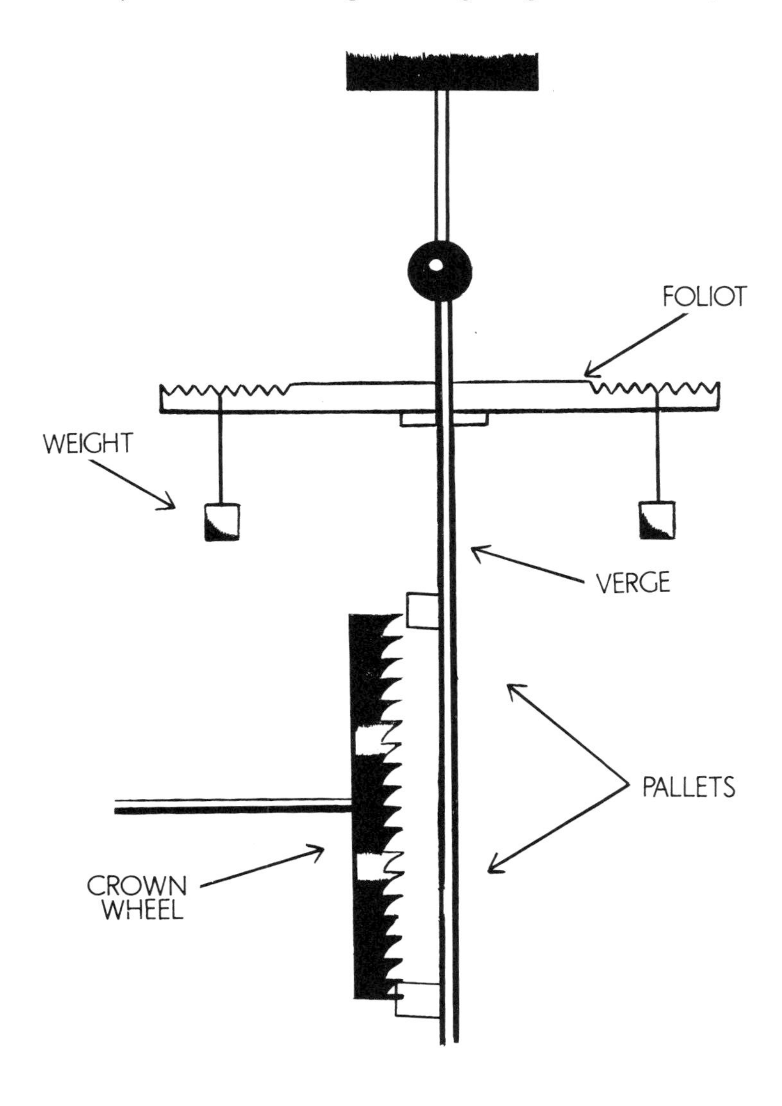

Foliot controlling the power from a falling weight

German, de Vic or de Vick, for the first time when he made a clock for Charles V of France.

MECHANICAL CLOCKS

After de Vic's clock had been made the idea of mechanical clocks spread across Europe, although little is known of their early history. The first ones about which we have definite evidence were in use during the first half of the fourteenth century. They were public clocks, but we have no information about the many experimental clocks which must have been made before success was achieved. Visconti set up a clock on the church of Beata Vergini (now called San Gottardo) in Milan in 1335 and something is known of three others in Italy before 1350. The Milan clock struck a bell every hour up to twenty-four. In 1344 Jacopo Dondi took the mechanical clock a stage further. He erected in the tower of the Carrara Palace, Padua, Italy, a clock which not only struck the hours but also showed the phases of the moon and other astronomical information. The earliest known mechanism for striking the quarters is that of Rouen's great clock, dated 1389.

The first striking clock made in England was about 1370. All such clocks were for public use, because they were far too large, too expensive and too few to be owned by individuals. Throughout England, towers were built on town halls, churches and other town buildings to take these clocks which, in the main, only struck the hours. The oldest clocks still in existence are believed to be at Rouen, and at Salisbury and Wells, England. The clock at Wells is worth seeing, for on the dial can be found the hours of the day and the age and phase of the moon. When the hours are struck a number of moving figures on horseback are set in motion.

These figures are called *jacks,* the word coming from 'Jaccomarchiadus', which means 'the man in the suit of armour'. Jacks were popular in the fifteenth and sixteenth centuries and their purpose was to strike the hours on bells before the days when all clocks were fitted with dials. Many interesting, intricate and well-known jacks can still be seen today, for example in Exeter, Oxford, Leicester, Wells and York in England and in St Mark's Square, Venice, Italy (see frontispiece).

The oldest extant English clock is probably that in Salisbury Cathedral. About 1386 it was installed in a bell-tower in the Close, but it is now in the North Transept. Like de Vic's, the Salisbury turret clock was originally controlled by a verge escapement and foliot balance. At some time it was altered to use a pendulum. No dial has been found and it is believed that this clock, which possesses a striking mechanism in addition to a going mechanism, was intended for striking the hours.

Until the invention of the mechanical clock, it had been the custom to divide day and night into the same number of hours. The English system divided them into twelve hours each, but counted twelve twice, from noon and again from midnight. Later, however, the day and night together were divided into twenty-four equal hours. In Nuremburg the day and night were divided into eight hours and twelve hours according to the season. In the Germanisches National Museum in Nuremburg there is a clock the dial of which is arranged to accommodate the changing number of hours. For many years the Italians used the twenty-four hour system and their clocks struck from one to twenty-four starting from sunset. Other parts of Europe kept to the two-by-twelve system counting from noon and midnight. This method was used in the Salisbury clock, and it is interesting to note that somewhere towards the end of the fourteenth century the expressions *before noon* and *after noon* became common usage. It is not an unusual sight in England today to see public clocks in church or cathedral towers.

These public, or turret, clocks were made of wrought iron and the work was done by a blacksmith. The clocks did not keep time more accurately than to about fifteen minutes a day, but the public quickly learned the value of time. Many well-to-do people even bequeathed sums of money for the erection and maintenance of public clocks. In 1463 in the will of John Baret of St Edmunds Bury, Suffolk, England (now Bury St Edmunds), the following bequest was made:

> *I will yeve and be qweth yearly to the Sexteyn of Seynt Marie church viij. s to keep the clokke, take hede to the chymes, wynde up the peys and the plummys as ofte as nede is, so that the seid chymes fayle not to goo thorough the defawte of the seid sexteyn who so be for the tyme.*

Mechanism of Wells Cathedral clock, 14th century, one of the oldest clocks in existence

Perhaps it was because of the inaccuracy of the early clocks that only one hand was fitted, which indicated the hours and quarters. Not until the middle of the seventeenth century, when time-keeping became more accurate, did minute hands appear as standard fixtures. Some of the earlier hands were extremely ornate. Many devices were tried during the next century to make the minute hand accurate and to avoid confusion between the minute and hour hands. The following are three of the methods tried:

1. A revolving dial was used with a fixed pointer.
2. Warriors with swords were used as pointers.
3. Changing figures corresponding to the current hour were shown through an aperture in a fixed dial.

The concentric minute hand was introduced about 1670. One of the largest minute hands, three feet wide at its centre and fourteen feet long, is fitted to the largest electric clock in the world, in Liverpool, England.

DIALS

We know little about exterior dials and documentary information is scarce. The oldest illustrations of exterior clock dials are probably those of an old church of Amsterdam, Holland, dated 1544, showing four dials, and a view of Innsbruck, Austria, dated 1577, showing dials on two different buildings. It is recorded that in 1344 the Dean and Chapter of St Paul's Cathedral, London, made a contract with Walter the Orgoner (or organ-player) of Southwark to supply and fix a dial for the clock. In the contract a roof is mentioned and this suggests an exterior dial. The handbook of the National Museum of Wales quotes a poem written about 1355 which contains the line,

'Woe to the blackfaced clock which woke me'.

Whether the poet was cursing the clock or describing its black dial we do not know. Dr C. F. C. Beeson in his recent study *Clockmaking in Oxfordshire 1400–1850* writes that he found a reference to an exterior dial in Oxford as early as 1505.

2
Mechanical Clocks to Chronometers

Not long after the invention of the mechanical clock there appeared the weight-driven domestic or chamber clock. Where the first one was made is not known, and in fact, there is less information about these clocks than the public ones. They were rare and in the early fifteenth century were bought by kings and princes only. At first wrought iron was used in their construction, but about the middle of the sixteenth century brass began to be widely used. It had a great advantage over iron, for it allowed of much finer working and consequently smaller and more accurate clocks. Brass was first used for making clocks, it is recorded, by the Italian Dondi, who said that his clock, completed in 1364 after sixteen years of work, was entirely of brass and bronze; but two hundred years were to pass before brass was again used in clock movements. Of course, Dondi was a scientist rather than a clockmaker and would have chosen what he thought best. Clockmakers, descendants of the clockmaking blacksmiths, may have been suspicious of brass and they continued to use iron. They were the technological descendants of the blacksmiths and were used to iron. Clock and case were fastened by wedges and other devices until the invention of the screw in the middle of the sixteenth century.

The domestic clocks were built into an open four-posted frame and usually the going and striking trains were placed one behind the other, the dial being in front. A third train or mechanism and weight would be fitted if the clock had an alarm. A further elaboration was an extra train for striking the quarters. Great care was taken over the construction of these neatly shaped clocks, and often the dials and supporters were splendidly decorated. The sides, however, were open and the mechanism exposed to view.

Because of its weight the domestic clock could not be

16th century iron domestic clock, front and side view

carried about the house from room to room, and it was therefore normally hung in the hall. When a sufficiently loud bell was fitted it could be heard all over the house. Most of the early clocks in use in England were of European make and it was not until the sixteenth century that clockmaking became an accepted art. Up to the seventeenth century the blacksmiths and locksmiths were sent for when a town wanted a public clock and, with the simple tools of their trade, they set to work. The clock was made from hand-forged iron; the teeth of every wheel were individually filed down by hand, as were the arbors or axles. The making of a clock was a laborious task which could take years and they were therefore extremely expensive. As shown by the exhibits in the Science Museum of Kensington, London, the blacksmith produced a clock that worked reasonably well, and even if it did lose or gain a quarter of an hour a day, that probably mattered little in a narrow and slow-moving world. On the Continent men specialised in clockmaking and quickly brought their work to a fine art with a high degree of accuracy. The small domestic clock, however, was beyond the capabilities of the blacksmith, excellent craftsman though he was, and thus England fell behind in clockmaking.

The English locksmith tried to carry on where the blacksmith left off, but he too was unable to compete with the highly skilled French, German and Italian clockmakers. In particular the Liechti family of Winterthus, Switzerland, became famous for their chamber clocks. Religious persecution did much to establish the clockmaking tradition in England, for many foreigners, clockmakers among them, fled from their own lands to settle there. The proficiency of these newcomers obviously worried the London apprentices, who revolted against them in 1517. So highly thought of were these foreign craftsmen that one, a Bavarian named Gratzer, was appointed *Deviser of the King's Horologies*. This post was undoubtedly the result of Henry VIII's keen interest in time-keeping and astronomy, and of the fact that his large collection of clocks needed periodic attention.

Protection for clockmakers had indeed been implemented long before Henry's time. In 1368 Edward III had granted a charter of protection for 'three Orologiers', all from Lombardy but one of whom had originally come from Delft.

Drum clock with detachable alarm, about 1600

These three nameless refugees worked on a great clock for Windsor Castle in 1353 and may also have done work for the King at Westminster, Queenborough and Langley.

As trade improved and the country became richer in

Elizabethan times, the demand for clocks, both large and small, increased. The English-made domestic clock differed in some respects from the foreign one. The mechanism was completely enclosed and the alarm bell was fitted above the top of the frame, making a well-balanced and neat ornament as well as a clock. By the end of the seventeenth century quite elaborate clocks, some of which were musical, were being made. Chamber clocks had a new form of regulator called a balance wheel. A large wheel was fitted to the verge and took the place of the foliot, cross-bar and weights. If the clock was required to go faster or slower the owner fitted a heavier weight or lighter weight and a clockmaker adjusted the weight of the balance wheel.

SMALLER CLOCKS

As clocks became more accurate and travel more common a need arose for an even smaller clock : a clock which could be carried about with ease. In the first decade of the sixteenth century a Nuremburg locksmith, Peter Hele, or Henlein, had produced a clock which was driven by a spring – the first true watch. These early 'self-going horologia' were spherical, but soon became drum-like in shape, being about six inches in diameter and but a few inches high. The flat face of the watch was protected by a curved metal cover pierced with holes so that the numbers could be seen. During the latter half of the century the Germans produced some fine and elaborate spring-driven clocks and watches. Even these smaller watches and clocks were fitted with minute hands, and occasionally second hands.

It was, however, soon discovered that spring-driven clocks and watches had one great disadvantage. A tightly coiled spring exerts a much greater force (thus making the clock go more quickly) than when it is in its normal loosely coiled position. A German invention called a *stackfreed* overcame this disadvantage. It resisted the power of the coiled spring but assisted it when it was running down. A more successful invention, the *fusee*, rapidly superseded the stackfreed.

In 1525 a Czech, working upon the drawings of the famous Leonardo da Vinci, produced a method of obtaining a uniform torque no matter what force was being exerted by the spring. It was a most ingenious device yet simple in construction. One end of a length of chain (steel wire can be used) is anchored to the main spring and the other wound round what looks like a miniature helter-skelter. When the spring is wound up it pulls the chain from the smallest groove in the fusee with ease. But as it unwinds so the leverage required to uncoil the chain from the grooves of larger diameter increases. Thus there is a uniform pull on the axis of the fusee which drives the mechanism of the clock as the spring runs down.

The fusee and spring replaced the weights in the domestic clock so that they no longer had to be hung but could stand on a shelf or mantelpiece.

The invention of the pendulum in the seventeenth century made timekeeping more accurate. The science of pendulum making and the uses to which the pendulum has been put are reputed to have arisen out of the observations of one man. Towards the end of the sixteenth century, so it is said, Galileo stood one day in the cathedral of Pisa in Italy and watched a lamp swinging to and fro. He noticed that although the swings of the lamp became shorter their time of swing remained the same. (A pendulum swings uniformly owing to the gravitational pull of the earth.) The pendulum was started by hand and only occasional impulses were required to keep it swinging. It was the Dutchman Huygens who, on Christmas Day in 1656, first patented and described a pendulum clock which he had designed. He had found a way of keeping the pendulum swinging by connecting it to the driving train of the clock. Thus, while the pendulum was kept swinging it also controlled the rate at which the wheels of the clock rotated, and the spring recorded the hours, minutes and seconds on the dial as well as providing the necessary 'kick back'. Adjustment was now a simple matter, for it is the length of a pendulum which determines the time of its swing. If a clock is too slow, the bob, or weight, on the pendulum is raised; if it is too fast the bob is lowered. On modern clocks this is usually done by means of

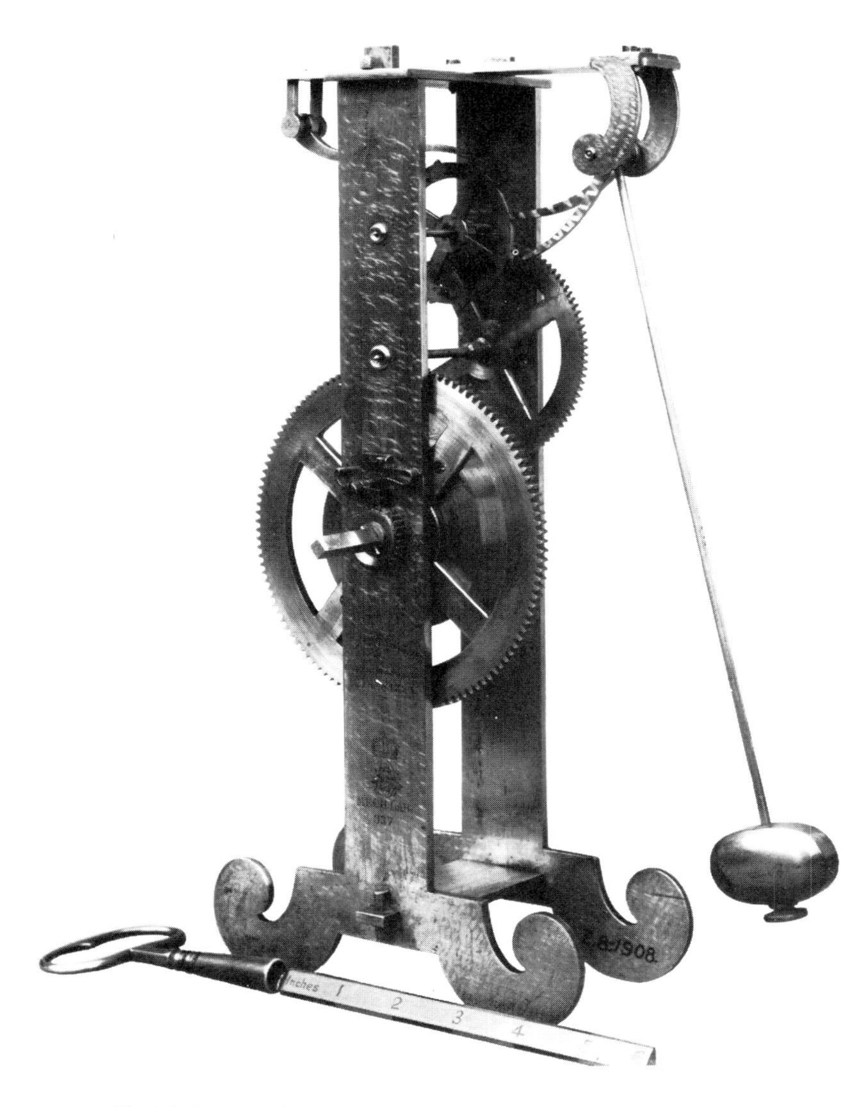

Model showing the application of the pendulum made by Galileo, 16th century

a slight turn of the screw which is beneath, or set in the bob.

Although the inclusion of the pendulum in clocks meant improved timekeeping, early clockmakers had realised that temperature upset their calculations. In his *Horological Dialogues*, 1675, J. Smith raised this problem in connection with the foliot, cross-bar and weights and the method of rectifying any errors:

> *These movements going with weights must be brought to keep true time by adding to or diminishing from them; if they go too slow you must add thin shifts of lead to the weights to make it go faster, but if it go too fast then you must diminish the weight to make it go slower; for that whensoever you find either to gain or lose, you must thus, by adding or diminishing, rectifie its motion; note that these Balance movements are exceedingly subject to be altered by the change of weather, and therefore are most commonly very troublesome to keep to a true time.*

And so it was with the pendulum. Adjustment of the screw, as already mentioned, is a simple matter, but there is still the question of expansion and contraction to be counteracted. Many methods have been used for compensating expansion, and the whole study of pendulums alone would fill a volume. The greater efficiency of the pendulum was realised at once and pendulums were fitted to most new clocks – and some old ones – during the latter half of the seventeenth century. Whether the early pendulum clocks were weight-driven or spring-driven the pendulum swing was quite a wide arc. Furthermore, the verge escapement checked its motion, and considerable thought was given by clockmakers to try to overcome this. The solution was finally produced, probably by Dr Hooke in 1676, and first used in that year by William Clement, a London watchmaker. This invention was the anchor or recoil escapement.

But first a few words about this remarkable inventor. Dr Robert Hooke was one of the great personalities of his age and his interest covered a wide field. He was a fine architect and his name is associated with Sir Christopher Wren's in connection with the construction of the Royal College of Physicians in London. Hooke must have given considerable thought to

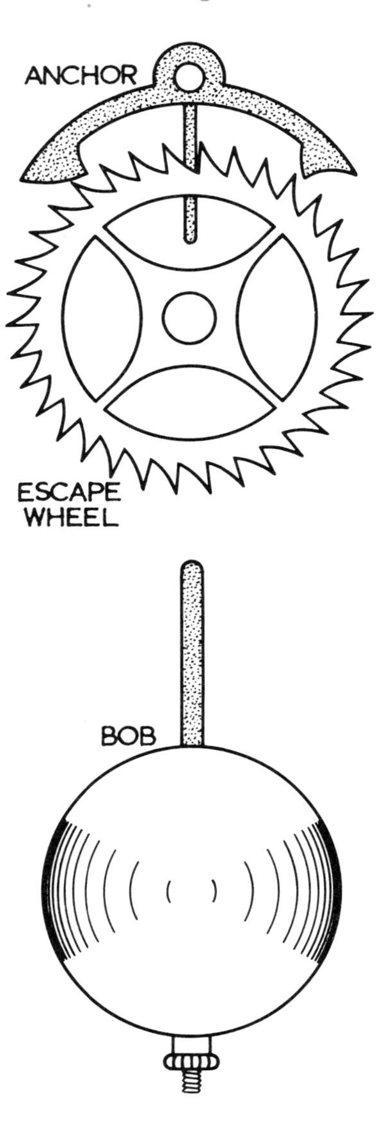

Anchor escapement and bob

the Great Fire of London of 1666 for he invented a fire engine. Another product of his inventive brain was a machine for cutting wheels. This was, of course, a great help to the clockmaker for until this invention the teeth of wheels were cut individually by hand. He is also credited with having invented the balance or hair spring, and of having carried out many important experiments with the telescope. There seems to have been time left over for the writing of a number of scientific books and for taking an active part in the astronomical field.

Living at the same time was Flamsteed the first Astronomer Royal, with whom Hooke probably corrresponded. Among his work Flamsteed catalogued the fixed stars and provided Newton with useful information. Later, James Brindley, another Astronomer Royal, discovered the nutation of the earth's axis. That is, he discovered that the earth has a wobble and takes nineteen years to wobble from one side to the other – in addition to its axial rotation and journey round the sun.

Among Hooke's friends were Sir Isaac Newton, the great philosopher and scientist, and Thomas Tompion, a famous English clockmaker.

Hooke showed how a long pendulum could be kept swinging by the push, or impulse, given by a pin in the rim of the balance. The escape wheel was set in the vertical plane and an arm like an anchor engaged its teeth. A longer pendulum than had been used before oscillated the anchor, and the escape wheel turned half the space of a tooth at every tick of the pendulum. Before the use of this form of escapement, pendulums were short and carried light-weight bobs, and had been used in both weight and spring clocks. With the new escapement, longer pendulums with heavier bobs could be used, and it was found that there was a far greater control over the timekeeping. In addition, the longer pendulum required a smaller arc.

With greater efficiency in timekeeping and a new overall shape in the mechanism, new case designs began to appear. Clocks were made to run for longer periods – a month or more – by the use of larger mechanism. The cost and increased size resulted in the three-day clock becoming the most common.

Before we leave the pendulum it is worth recording that it was not until the middle of the eighteenth century that an ingenious yet simple method of compensation was tried out successfully. Wooden pendulums with lead bobs were made. In hot weather the wood expanded, thus lowering the centre of oscillation, that is, the control point of the pendulum swing. The lead bob resting on a screw could only expand upwards, thereby counteracting the expansion of the wood.

THE CLOCK REPAIRERS

Clocks needed repairing in earlier centuries as they do today and there are records to show that particularly in the Middle Ages the repairers were kept busy on the public clocks. Roger the Clockmaker was sent from Barnstaple to Exeter Cathedral

5.

HOROLOGIA FERREA.

Rota æqua ferrea ætherisq; voluitur, Recludit æquè et hæc et illa tempora.

An early clockmaker's business

in 1424; the 'clokkemaker of Kolchester' repaired a clock for the Duke of Norfolk about 1483. At Rye, Sussex, in 1533 a Frenchman was employed to repair the church clock; a native of Gascony, one Lewis Billiard, supplied a new one in 1561.

Elizabeth I was a keen clock and watch collector and had her own clockmaker or clockmender, one Nicholas Urseau, of French descent. In addition the Queen employed a clock-keeper, Bartholomew Newsam, who succeeded Urseau in 1590 on the latter's death. Newsam, a Yorkshireman, became the first English Royal Clockmaker. In the British Museum there is an example of his fine workmanship. Randolph Bull, who later became the Royal Clockmaker, was employed to make the clockwork for Thomas Dallam's organ in 1599, a gift from the Queen to the Sultan of Turkey.

Many of these earlier repairers had been foreign or of foreign descent but gradually the craft of clockmaking and repairing became established in this country.

THE CLOCKMAKERS' COMPANY

In 1630 the Clockmakers' Company was formed. Arms were granted to the Worshipful Company of Clockmakers forty-one years later, their motto being *Tempus rerum imperator*, or 'Time is the ruler of all things'. This Company united all the clockmakers and watchmakers, whether English or foreign, and they had to abide by its rules. Clocks became more common and cheaper, but even so they were still too dear for any but the well-to-do households. The church clock still remained, for some time to come, the only one available for most people. The men who made the various parts of the clocks and watches – spring makers, gilders, clockcase engravers – also became members of the Company.

The function of the Company was to protect the English clockmaker. It also controlled the number of new entrants to the trade, trained them and kept an eye on the quantity and quality of clock production. The right of search of members' premises by officers of the Company ensured good workmanship but, it appears, they were unable to control the number of apprentices whom London clockmakers employed. Clocks of poor quality were ordered to be destroyed. In its early days the Company was not well off and had no large hall of its own, members having to hold their meetings in various taverns; but over the past three centuries the position has improved. Among the Company's members have been some of the finest craftsmen in London.

THE LONG CASE CLOCK

The long case clock was first made in England in 1658. Much later it became known as a Grandfather Clock, and it was probably so named after a popular song of the 1880s, which went:

My grandfather's clock was too big for the shelf,
So it stood ninety years on the floor.

With the coming of the pendulum it quickly became the most popular of all weight-driven domestic clocks. It has altered during its three hundred odd years and is still popular today.

In London, at sea level, a seconds pendulum must for accuracy be 39·1 inches long, at the equator 39 inches, and in Edinburgh almost 39·11 inches. The required length varies with altitude also, but it was agreed to use 39·1 inches as the accepted length for long case pendulums. The design of these clocks, though regulated by the length and swing of the pendulum, became almost as important as the clock mechanism itself. The case was made of wood and invariably highly polished and richly ornamented. The dial was enlarged and was just as richly decorated as the case. The mechanism was enlarged also to include bells of varying sizes which gave rich and different tones for the striking of the hours and quarter hours. It was not long before extra dials and mechanism were incorporated in the long case clocks, and they then automatically registered the month, date of the month and day of the week. Early in the eighteenth century further additions were made to show the course of the planets, the position of the sun and the moon and the rising and setting of any fixed stars. Thomas Tompion made one of these clocks, and with his assistant and nephew by marriage, George Graham, was responsible for some remarkable clocks and very fine workmanship.

A good London-made grandfather clock towards the end of the seventeenth century with all its fine carving, gilt metal mountings, highly polished walnut and ebony would have cost about £24 ($56.80) or, in modern currency representing the same value, well over £150 (about $360). By the beginning of the next century lacquered cases largely replaced the walnut and ebony cases, such was the change in fashion.

A further addition was made by English clockmakers towards the end of the seventeenth century when a dial was incorporated in the clock to show the equation of time; that is, the difference between clock time (Mean Time) and time by the sun (Solar Time, or apparent time). Only four times a year do these two times coincide, and as the clocks were set to the mean time by the sundial, it was helpful to know always the difference between apparent and mean time. Printed equation tables were published, and the one illustrated was printed for Tompion in 1690.

REFINEMENTS

The eighteenth century produced a rather different type of clockmaker. He still made very elaborate clocks and kept to the accepted shape of the long case clock, but he gradually simplified the mechanism. Further, and more important, the exploitation of earlier inventions resulted in even greater accuracy being achieved. George Graham (1673–1751) made two important additions to the clockmaker's craft – the dead-beat escapement and the mercurial compensation pendulum.

In order to maintain the balance of a clock or the vibrations (or swing) of the pendulum, the escapement must transmit impulses of the motive power. The escapement controls the speed of the revolution of the train when the escape wheel is at rest. The manner in which this is done varies with each different kind of escapement. This control of the motive power operates when the balance or pendulum is oscillating or vibrating.

TABULA

Æquationum Dierum Naturalium

Temporis Intervalla

Inter Horas ab *Horologiis Oſcillatoriis* æquabiliter moventibus, & à *Scioterícis Solaribus* accuratis indicatas, ad Diem quemlibet ſingulum Anni 1690. St. Novi, exhibens: Futuris etiam Annis hujus ſæculi, abſque ſenſibili differentiâ inſerviens.

Dies	*Januar.* Mi.	Sec.	*Februa.* Mi.	Sec.	*Mart.* Mi.	Sec.	*April.* Mi.	Se	*Maii* Mi.	Se.	*Junii* Mi.	Se.	*Julii* Mi.	Se	*Aug.* Mi.	Se.	*Sept.* Mi.	Se.	*Octob.* Mi.	Sec.	*Novem.* Mi.	Sec	*Decemb.* Mi.	Sec.
1	4	42	14	20	12	37	3	44	3	20	2	50	3	04	5	35	0	31	10	30	16	01	10	02
2	5	09	14	27	12	25	3	26	3	28	2	41	3	16	5	31	0	50	10	48	16	00	9	38
3	5	36	14	32	12	12	3	08	3	35	2	31	3	27	5	27	1	09	11	06	15	59	9	14
4	6	04	14	37	11	57	2	49	3	42	2	21	3	38	5	22	1	29	11	24	15	57	8	49
5	6	30	14	41	11	43	2	31	3	47	2	11	3	48	5	16	1	49	11	41	15	55	8	23
6	6	57	14	44	11	28	2	13	3	52	2	01	3	59	5	10	2	07	11	58	15	52	7	57
7	7	23	14	47	11	13	1	55	3	56	1	51	4	09	5	03	2	27	12	14	15	48	7	31
8	7	48	14	48	10	58	1	38	4	00	1	40	4	18	4	56	2	47	12	29	15	43	7	05
9	8	12	14	49	10	42	1	21	4	03	1	28	4	27	4	48	3	08	12	45	15	37	6	38
10	8	35	14	49	10	25	1	05	4	06	1	17	4	35	4	39	3	28	13	00	15	31	6	10
11	8	59	14	48	10	08	0	49	4	09	1	05	4	43	4	30	3	48	13	14	15	23	5	42
12	9	21	14	47	9	51	0	32	4	11	0	53	4	51	4	20	4	09	13	28	15	15	5	13
13	9	43	14	45	9	34	0	16	4	12	0	40	4	58	4	10	4	29	13	42	15	05	4	45
14	10	05	14	49	9	17	0	*01	4	13	0	28	5	05	4	00	4	50	13	56	14	55	4	16
15	10	26	14	38	8	59	0	14	4	12	0	16	5	11	3	49	5	10	14	08	14	44	3	47
16	10	45	14	33	8	42	0	29	4	11	0	03	5	17	3	37	5	31	14	20	14	32	3	17
17	11	04	14	28	8	24	0	44	4	10	0	*10	5	23	3	24	5	51	14	31	14	20	2	47
18	11	23	14	23	8	05	0	58	4	08	0	23	5	28	3	11	6	12	14	41	14	06	2	17
19	11	40	14	14	7	47	1	12	4	06	0	36	5	32	2	58	6	33	14	51	13	52	1	47
20	11	57	14	09	7	29	1	26	4	04	0	49	5	35	2	44	6	53	15	01	13	38	1	18
21	12	14	14	01	7	10	1	38	4	01	1	01	5	38	2	30	7	14	15	11	13	21	0	48
22	12	30	13	52	6	52	1	51	3	57	1	14	5	41	2	16	7	34	15	20	13	04	0	18
23	12	44	13	43	6	33	2	03	3	52	1	27	5	43	2	02	4	54	15	26	12	47	0	*12
24	12	58	13	34	6	15	2	14	3	47	1	40	5	45	1	46	8	14	15	32	12	28	0	42
25	13	12	13	24	5	56	2	24	3	41	1	53	5	46	1	30	8	33	15	38	12	09	1	12
26	13	24	13	13	5	37	2	35	3	35	2	05	5	46	1	13	8	53	15	44	11	50	1	42
27	13	35	13	02	5	18	2	46	3	29	2	17	5	45	0	56	9	13	15	49	11	30	2	11
28	13	46	12	50	5	00	2	56	3	23	2	29	5	44	0	39	9	32	15	52	11	09	2	40
29	13	56			4	41	3	04	3	15	2	41	5	42	0	21	9	52	15	55	10	47	3	09
30	14	05			4	22	3	12	3	07	2	53	5	40	0	03	10	11	15	57	10	25	3	38
31	14	13			4	03			2	58			5	38	0	*14			15	59			4	07

OBſervato accuratè tempore apparente, Scioterici ope vel alterius alicujus inſtrumenti Mathematici, capiatur è Tabulâ præcedente temporis Æquatio diei competens, & diſponantur indiculi Horologii tardiùs aut velociùs tempore apparente (juxta Æquationis in Tabula Titulum) quantum præcisè ea eſt: Tunc ſi motu æquabili moveatur Horologium, differentiæ temporum ab eo, & à Scioterico commonſtratorum, Temporis Æquationibus, *obſervationum diebus*, in Tabula adſcriptis, ſemper deinceps æquales erunt.

LONDINI:

Sumptibus *Thomæ Tompion*, Automatopœi, ad Inſigne *Horologii* & *trium Coronarum* in Vico vulgò dicto Fleetſtreet, 1690.

Equation table of apparent and mean time printed for Thomas Tompion, the English clockmaker, in 1690

It had been noticed that in the existing recoil escapement, the recoil, by reason of its pressure and rubbing action throughout the train, added a little to the friction and wear on the gear teeth, pinions and pivots. The extra friction is negligible, and because of its robustness this type of escapement is cheap and is still the most widely used. The dead-beat escapement gives better results and does not wear so readily as the recoil, but it has to be accurately made and therefore costs more. Graham's invention stood for nearly two hundred years as the closest mechanical timekeeper known to science, and a number of his clocks still exist, and work, in famous observatories.

The dead-beat escapement, so called because the train is stopped dead, without trace of recoil, at each beat of the pendulum, was the first to give results of high accuracy. The train is stopped by a tooth of the escape wheel dropping on to a 'dead' face of the pallet. As the dead face is a curve forming part of the pallet face the tooth merely rests while the pendulum finishes its swing. Just at the end of the swing the tooth slips off the corner of the pallet and by pressing against the impulse face of the pallet gives a light push which keeps the pendulum in motion. The wheel turns forward and another tooth hits the other pallet and locks on it until it is released to impulse as before. If correctly adjusted for beat and timekeeping, a pendulum clock regularly wound need never stop except when cleaned and oiled.

Various types of escapement may be used with a balance wheel, though now the lever escapement is common. It is called the lever escapement because the pallets in contact with the escape wheel are pivoted on a small lever. The escape wheel is fixed on to the last pinion of the train and the shape of the teeth varies according to whether the impulse plane is on the tooth or the lever, or partially on both.

There are three main kinds:

1. Club teeth – special shaped escape wheel teeth employed with the lever movement to increase the length of the impulse but at the same time decreasing the wear.

2. Pointed teeth – called the English lever escapement.

3. The shape having the club teeth and the pallet in the form of a pin, a method used in some pocket and wrist watches (the Swiss Roskopf type).

Brass or steel are usually used for the escape wheels and the number of teeth varies from twelve to eighteen. The average is fifteen.

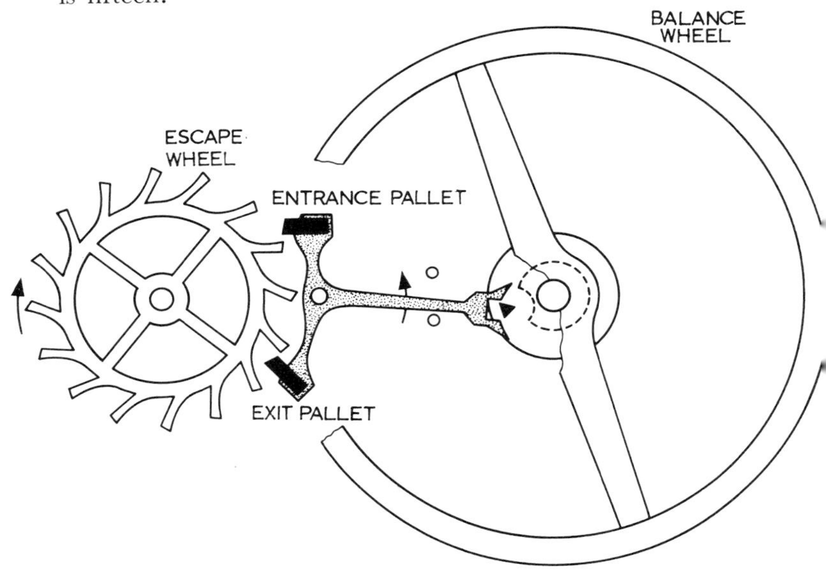

Lever escapement

The lever is formed by two arms, slotted to hold the pallets. At the opposite end of the pallets the lever is terminated by a fork connected to the lever by a rod. This fork has an entrance into which the impulse pin enters to unlock, and so receive the impulse. It has a stud into which the safety guard pin is fitted. The lever is of steel or brass and the pallets are invariably of hard stone such as synthetic corundum (ruby or sapphire) or garnet.

The double roller is also of brass or steel and is made of two discs joined together by a tube or collet. The upper disc is called the impulse roller and carries the impulse pin. The lower disc is the safety roller.

The most common escapements are:

the anchor – for pendulum clocks;
the lever – for clocks and watches; and
the detent – for marine chronometers.

There are four categories of escapement:

1. The frictional escapement. Here the escape wheel is in contact with the balance staff or a part of it during the whole oscillation of the balance.

Example of frictional escapement may be found in cylinder escapement for watches; Duplex and Virgule escapement for watches; Graham's dead-beat escapement for clocks; recoil escapement for clocks; verge escapement for clocks and watches; Brocot's escapement for clocks.

2. In the free escape category there is an arrangement whereby the escape teeth are arrested and then made free. The balance can thus rotate in an arc of oscillation in complete freedom. This method is used in lever escapement for watches and Riefler's anchor escapement for clocks.

3. The Robin escapement, which is a combination of lever and the detent and is used for watches.

4. The rotating escapement by Karrusel, which has a mechanism allowing the rotation of the complete escapement.

The escapement is a method used to stop and transmit motive power. To do this it is necessary to have a system of control to unlock the escape wheel before the balance can receive an impulse. This unlocking disturbs the timing and efforts by inventors to minimise this have gone on for centuries.

Graham's other invention, the mercurial pendulum, was the most accurate timekeeper up to this century and was a further improvement in the field of compensatory experiments.

Until the middle of the eighteenth century practically no attention had been paid to the oiling and care of clock mechanism. Indeed, little had been done to exclude dust and dirt from entering the clock. It eventually dawned upon the clockmakers, however, that cleanliness was essential if clocks were to be accurate, and soon clock mechanism became totally enclosed. It was also realised that continual friction of the trains resulted in their becoming loose. In 1704 Debaufre and Facio experimented with rubies set in the holes in which the arbors revolved. Rubies, being harder than the metal, produced less frictional wear and so greater accuracy. This did not become a common method of overcoming wear and friction for another century.

During the early part of the 1700s the musical box increased in popularity. It played a tune at each hour, the music being produced as in the modern musical box by a pin barrel. A pin barrel is a cylinder from which protrude a number of small pins. As the cylinder revolves, a pin touches a piece of spring, which then emits a pre-determined musical note. By setting the pins in a given order during manufacture and thus causing the notes to be sounded in order, a simple tune can be obtained.

Watches, too, had greatly improved. There had been some abolition of the fusee in favour of the cylinder escapement and the Lepine calibre. However, the fusee was retained in English lever for it continued to have a good timekeeping characteristic, and the English lever proved to be a better system than the cylinder escapement. Watches became thinner and less bulky without any loss of efficiency. Minute hands, and later, second hands, appeared in the first half of the eighteenth century.

During the same century there were many inventions improving timekeeping (some mentioned earlier in connection with escapement), but although the clockmaker continued to make as fine movements as they had ever done there was generally a lack of high quality in the dials. The demand for less expensive clocks of the bracket and tall case types resulted in the earlier silvered hour ring on a fine surface being replaced by a simple painted or enamel dial. Finely made spandrels or corner ornaments for dials were a feature of this period, the winged cherub being a popular motif with bracket clocks.

Clock- and watch-making spread and important centres in England sprang up at Leicester, Liverpool, Derby, Bristol, Newcastle-upon-Tyne and Coventry. However, away from London the emphasis appears to have been upon the weight-driven clock rather than the spring-driven type. This was because the specialist craft of spring making was still carried on in London.

By the beginning of the nineteenth century new designs in table clocks were appearing. Smaller clocks were being made for the narrow chimney shelf of that period, and the whole craft industry of watch- and clock-making was gradually being replaced by the factory system of manufacture. Although this did not reach the same proportion as in Switzerland and America, there were sufficient factory-produced clocks and watches to lessen the feeling of pride and craftsmanship in the handmade models.

Also, cheaper foreign watches were allowed into Britain duty-free and the more expensive British products suffered as a consequence. The only advantage of the mass-produced watches and clocks was to reduce the price and thus put these instruments into the hands of a wider and ever increasing population.

The period between the end of the eighteenth century and the middle of the twentieth century has been, in the main, one of steady improvement, but with few startling inventions such as had marked the earlier days of clock-making. Fascinating and strange kinds of watches and clocks continued to appear. One of the most interesting was the mystery clock invented by Schmidt of London in 1808. One variety of this clock showed two large hands freely attached to nothing but a sheet of glass, which apparently worked on their own. In fact, each hand contained a watch movement in its counterpoise which altered the centre of gravity as it ran and caused the hand to take up a new position.

As we have said, accuracy was the main aim; and between 1750 and 1850 clockmakers began to pay much greater attention to accurate gearing, transmission and the problems raised by friction. During this period a number of clockmakers became famous for a wide range of contributions.

Mention has already been made of Facio's solution to the friction problem, but in 1754 a further improvement in accuracy was reached. Thomas Mudge, second son of an English clergyman, invented the detached lever escapement and this invention is recognised as the greatest single improvement, except the balance spring, to be applied to watches. Yet it remained neglected for a century. The first watch to be fitted with this escapement was made by Mudge for King George III and it may still be seen at Windsor Castle. Later, Mudge was appointed clockmaker to the King and it is strange that he made no attempt to claim the invention as his own.

In 1770, John Arnold (1730–99), whose father was a Cornish watchmaker, invented what may be called the true chronometer when he introduced the compensation balance. Arnold's work was improved by Thomas Earnshaw (1749–1829) and the ship's chronometer used today is as Earnshaw designed it.

At Neuchâtel, Switzerland, in 1742, was born Abraham Louis Breguet, who was destined to rise to a high position in the list of horologists. At the age of fourteen he was apprenticed to a watchmaker at Versailles and later, in 1787, he started his own business. Breguet was a perfectionist and was not only a brilliant and ingenious horologist but introduced his own distinctive styling in cases, dials, hands and mechanism. Among his numerous inventions are the overcoil of the balance spring, the Tourbillon (where the whole of the escapement was enclosed in a cagelike structure and revolved once every minute), the shockproof watch and the free pendulum. At one time, working for Breguet, was another Swiss, Ingold. He invented a different form of lever escapement and a special and valuable tool called the Ingold fraise, with which the correct form could be given to the teeth of wheels.

Some inaccuracies in watches and clocks were bound to result as metal expanded with heat and contracted with cold. An Englishman John Ellicott (1706–72) gave this problem some thought and finally invented a compensation pendulum which employed the differential expansion of steel and brass to raise or lower the pendulum with a rise or fall in temperature. In France, the well-known horologist Pierre Le Roy invented the first detached escapement. This was an improvement on Mudge's escapement, which was not entirely free as the lever rubbed upon the roller of the balance.

CHRONOMETERS

There is another aspect of clock-making which should be included in this book, for the study of sea navigation presented clockmakers with a most intriguing problem. How could they produce an instrument that would give a ship's position anywhere at sea? There was considerable need for a solution to this problem in the eighteenth century owing to the rapidly increasing sea traffic. Various timekeepers made by such men as Huygens had been only partially successful.

It was necessary to compensate for the effects of heat and cold in any instrument made, and he who could do this satisfactorily would be well on the way to producing the required instrument. The Englishman Harrison finally found the solution. He knew that metals have a different rate of expansion and with this knowledge he made a bi-metallic compensation curb which is now used in thermostats. He riveted together a strip of brass and a strip of steel, and, when heated or cooled, the resulting bar bent. Compensation for the change of stiffness with temperature change, which affects the rate of a watch, was overcome by the use of the bi-metallic strip, the two metals enabling the effective length of the strips to vary.

In 1714 the British Government offered an award of £10,000 (about $25,000) for a satisfactory method of finding longitude to within an accuracy of one degree when a ship sailed to the West Indies and back. The prize was to be increased to £15,000 (about $37,500) if the calculations were accurate to forty minutes, and to £20,000 (about $50,000) if accurate to within thirty minutes. One degree of longitude is equal to four minutes of time, thus in order to win the whole award, a competitor had to make a timekeeper which did not err more than 30, or half a degree, by the end of the six weeks' voyage.

One of Mudge's tasks was to report to the Board of Longitude on Harrison's No 4 timepiece, which was concerned with a method of finding accurate longitude. Like Harrison, Mudge was determined to win one of the awards offered by the government, and he did in fact eventually receive £3,000 from a Committee of the House of Commons. Mudge, who had invented the detached lever escapement for watches, followed Harrison in designing marine timekeepers. Both his and Harrison's designs were tested at Greenwich but the mechanism made by both men was too complicated for general use. Early models produced by Harrison were not accurate enough, and Mudge died before his work was finished. However, Harrison succeeded in the end.

John Harrison, a Yorkshire carpenter, took up the challenge. He spent his life on the problem, and in his first chronometer he overcame the effects of heat and cold by varying the initial tension of the thin wires connecting the four balance springs to the balance arms. He connected his new bi-metallic strips in such a way that any temperature change provided sufficient power, transmitted through a series of levers, to increase or decrease the tension of the balance springs by shifting the position of their inner and fixed ends. Not satisfied with chronometer No 1 he made two more.

He used different arrangements for compensation in No 3. The balances were circular, connected together by wires as before and controlled by a single large spiral spring with one and a half turns, fitted to the staff of the upper balance. He devised an automatic way of regulation, by mounting the curb pins at one end of his bi-metallic strips while the other end was free to move. A rise in temperature would cause the strip to become convex on the brass side and concave on the steel side, and the curb pins would thus be moved along the balance spring. By adjusting the length of the curb, it could be so arranged that the amount of movement in the instrument's working would shorten the spring exactly enough to compensate for the retardation produced by any rise in temperature. A fall in temperature would lengthen the spring to produce the necessary effect in the opposite direction. The initial position of the curb pins could be adjusted by moving the framework carrying the curb so as to alter the instrument's rate of going.

Trials with this chronometer were successful but the Board

Harrison's number 4 chronometer. Only 5 inches in diameter, it has a single balance wheel and the escapement used is a much improved modification of the verge

of Admiralty were not satisfied that the instrument's good performance was not purely accidental. They knew nothing of the working of the chronometer; Harrison refused to tell them his secret. From time to time the Board had advanced sums of money to help him in his work and he was offered £2,500 ($6,250) for his 'useful invention', a sum of money which was to be deducted from any awards to which he might be entitled to later on as a result of further work.

In 1764 Harrison was permitted to carry out a second trial, and *HMS Tartar* sailed from Portsmouth for Jamaica, with his No 4 chronometer. The overall total error after 156 days was a gain of only fifty-four seconds while if allowances were made for the changes in the rate in different temperatures declared by Harrison before sailing, the error could be reduced to a loss of fifteen seconds in five months. This was less than one tenth of a second per day.

The Board had to accept this proof of the value of Harrison's chronometer but they offered him only half the award. The remainder would be exchanged for the secrets of his chronometer and when similar ones of his making were equally successful. He fought against this injustice but he finally gave in and received the award. Harrison had his followers and numerous improvements to his chronometer were made later.

3
Micro and Macro Mechanism

Harrison had shown that a timekeeper could be made which kept sufficiently accurate time at sea. However, the instrument was complicated and difficult to construct. Progress was rapid and by 1820 chronometers of the type used today were being produced in large numbers. The work became an important branch of the watch trade. For some seventy-four years before 1914 the Greenwich Observatory held annual trials for box chronometers, each trial lasting twenty-nine weeks.

The method of obtaining an order of merit, the award of what was known as the 'Trial Number', of those chronometers submitted was based upon a formula $(a + 2b)$, where 'a' was the algebraic difference between the greatest and least weekly sums of daily rates and 'b' the greatest difference between the sums of daily rates for two consecutive weeks. This system was evolved from a precise investigation of the Navy's needs, and the tests included two periods of four weeks each in the oven, in temperatures ranging from some 75° to 100°F, as well as three periods of seven weeks each in the room. There were no icebox tests. The average annual number of chronometers taking part over a period of ten years, including and preceding 1914, was approximately fourteen. The average of those successfully accurate during that time was about ten.

Thomas Mudge, who had invented the detached lever escapement for watches, quickly followed Harrison in the new branch of chronometer-making. He invented a constant force escapement which in theory was perfect, but in practice was not so successful. Unfortunately he died before being able to complete his investigations.

Pierre Le Roy, the French inventor of the first type of detached escapement where the balance wheel was free from any interference during much of its swing, also studied the problem of finding longitude. In 1766 he made an entirely original type of chronometer. In this instrument he incorporated his detached escapement and the first compensation balance. Compensation for changes of temperature was achieved by varying the effective size of the balance. Harrison, it will be recalled, had varied the effective length of the balance spring. At this time details of Harrison's No 4 chronometer had not been made public, and it is to Le Roy's credit that independently he should have made an instrument which, on test, was found to be the equal of Harrison's.

The spring detent form of chronometer escapement, now almost universally used in these instruments, was believed to have been invented by another Frenchman, Berthoud, but the Englishmen Arnold and Earnshaw made the same claim. What we do know is that the Frenchman knew his job and made some seventy chronometers, and that the Englishmen turned a highly skilled and individual craft into a commercial proposition. There were numerous experiments for making the time of swing of the balance independent of variations of the driving force and of the friction in the train wheels. Better compensation for the effect of temperature changes was continuously sought. Arnold studied the balance spring, for he was aware that in an ideal simple harmonic motion the period of an oscillation is entirely independent of the amplitude of swings, and swings of different amplitudes are said to be isochronous. The motion of a chronometer balance wheel controlled by its spring gives almost this ideal situation. Arnold's helical form of balance spring showed that the swings of the balance could be made almost perfectly isochronous provided the two ends of the balance spring were bent into the form of certain curves. However, only in 1861, after Phillips had carried out extensive experiments and produced a series of charts and reports, was it possible to find out the exact curves for helical springs for chronometers and spiral springs for watches.

An additional problem is that not only does a balance expand with a temperature rise, but at the same time its stiffness decreases. It was realised after experiments had been carried out that the effect of the change of stiffness is ten times as important as that of the expansion of the balance. The problem then was to design some form of compensation balance whose moment of inertia varied with temperature according to a law similar to that of the variation of stiffness of the balance spring. Many people tried to find the answer. Dr Guillaume of Paris produced an alloy called invar whose expansion with change of temperature is almost negligible. Later he produced another alloy to take the place of the steel bar of the bi-metallic strips then being used in the compensation balance. He found that this alloy, combined with brass, did the trick, and he used it in the Guillaume or integral balance. This balance is able to compensate almost perfectly for the change in stiffness of the steel balance spring with changing temperature. This was a big step forward. He continued with his researches and about 1920 made another alloy of nickel, steel, chromium and other elements, calling it elinvar. This alloy's elasticity was almost independent of a temperature change and was, furthermore, non-magnetic. So a balance spring of elinvar not only avoids magnetic disturbances but removes the effects of temperature changes. Chronometers made by the Swiss Ditisheim in the early 1920s, and incorporating Guillaume's elinvar balance springs, were highly successful. These methods are now used in high-grade watches and the Swiss, in particular, use the elinvar balance spring in many of their watches.

WATCHES

In a school book written by Johann Cocleus (1479–1522) of southern Germany, and published in 1512 in Nuremburg, there is a reference to his compatriot Peter Henlein and his clock-watch.

> *. . . a young man, Peter Henlein, has constructed works which amaze the most skilled mathematicians; for from pieces of iron he makes horologies containing many*

wheels; these horologies may be carried in any position, having no weights, and even in the pocket of the jerkin or in the pouch they go for forty hours and strike.

The first watch is supposed to have been made by Henlein about 1500. It was, as Cocleus wrote, a portable clock, so heavy that it had to be hung from a belt around the waist. In the 470 years since Henlein's day there have been great advances in watch-making and today an ordinary good watch keeps time to within half a minute a day; that is, its error of running is less than 0·04 per cent. Watches of such antiquity were spherical or drum-shaped, and six of the former shape have survived, one of them in Germany, made probably before 1550.

The earliest dated French watch, also spherical, was made by Jacques de la Garde about 1551. The drum-shaped watches, which finally superseded the spherical type, were usually enclosed in gilt or bronze cases. Some cases were of silver or gold and even such cases would be cheaper than the watch movement they enclosed. Because those early watches were too large for the pocket and had to be worn on a cord or chain, the cases were richly ornamented. From 1625 onwards English watches became plainer and this may have been a result of the growing Puritanical movement.

Almost all watches surviving from before 1590 are German. They differ widely from the French and even from the English which were derived from them. We should therefore first consider the German watches.

EARLY GERMAN WATCH-MAKING

Most of the sixteenth-century watches have either a striking mechanism or an alarm mechanism. Many watches had both and so the elaborate cases were pierced round the edge and on the bottom. Usually the bell was fixed to the bottom of the case and the movement itself hinged to the case. A disadvantage here was that it had to be turned out of the case to be wound up. Occasionally the bell was fixed to a hinged cover, and then the movement dropped into the case from above and was held in place by a latch. The dial in many instances had an outer ring of figures I–XII and an inner ring numbered 13–24. This was to satisfy Italian, Bohemian and south-west German timekeeping, which included the use of the later numbers. In all German watches of the sixteenth century the Arabic figure 2 is formed like a Z. Many watches and clocks of the period were provided with raised knobs at the hours to enable people to find the time during darkness. Usually that part of the dial inside the chapter ring was engraved with a rose or star-shaped pattern. This chapter ring was the circular metal plate which carried the numbers of the hours and later, as on a religieuse made by Hanet in Paris, about 1660–5, was numbered for every minute. The dial plates were either metal gilt or silver, but where enamel was used for decoration, then silver was always used as well. Watch hands were made of steel and most carefully shaped. Sometimes flat unchiselled and ugly hands are found on the early watches but such hands are invariably later replacements. There was no need to open up the watch to see the time for the dial cover was always adequately pierced. The Thirty Years' War (1618–48) dealt the German watch industry a severe blow from which it was never to recover. A vigorous French industry was to take the lead.

FRENCH WATCH-MAKING

The French industry goes back to the sixteenth century and, in fact, it is believed that watches were being made at Blois as early as 1518. The watch-making at Blois grew in prosperity and importance although minor centres were established at Autun, Rouen, Paris, Sedan, Lyons, La Rochelle, Dijon and Grenoble. Jacques de la Garde made a fine watch in 1551, but few other early examples of French watches have survived. In shape they are spherical or oval and none is of the circular drum type. From about 1590 French watches became fairly numerous. They were oval in shape and had straight sides, and the gilt metal cases were covered with engravings of figures, foliage and scenes of different kinds. The dial plates of these

17th century French oval alarm watch by André Pichon of Lyon: top, general view; bottom, the movement

French watches were usually gilt with engraved silver chapter rings. The French did not pierce their watch cases as did the Germans. Jean Valliers of Lyons made some thirteen splendid watches. Although there was nothing attractive about the cases, the superb movements, the workmanship of the striking, the alarum and calendar mechanisms and the fine decoration all show that Valliers was indeed one of the greatest craftsmen of the age.

ENGLISH WATCH-MAKING

The English were a long way behind the French and German watchmakers and we know of no English watches being made before about 1580. Indeed, only just before this period do we hear of Queen Elizabeth I's watchmakers, Urseau and Newsam. By the end of the century any English-made watches were simply copies of French and German products. Randolph Bull of London, working at the end of the sixteenth century, produced a good example of the composite watch. The case of this striking watch and the inner ring of 13 to 24 hours on the dial are German. However, the 2s are not formed like the German Zs, and there are no raised knobs to indicate the hours. The mechanism with its fusee is undoubtedly French.

During the early part of the seventeenth century English watch-making developed a style of its own and the workmanship was soon as good as the best in continental watches. The popular watch in the middle of the seventeenth century was known as the Puritan watch. It was a plain, oval one, about two and a half inches long. Although it was simple on the outside, the movement was highly decorated, and this type of watch may indeed have been carried in the pocket. Pockets in those days were confined to breeches and were not considered suitable for carrying watches. It is unlikely that you will see a watch in an early seventeenth-century painting, and not until about 1675, when Charles II made the long waistcoat fashionable, do we see watches worn in a pocket. English watchmakers seem to have been more concerned with the mechanism of timepieces than with the elegance of their cases. Pinchbeck, an amalgam of copper and zinc intended to imitate gold, was increasingly used for watch cases. The emphasis upon horology rather than decoration resulted in the acknowledged supremacy of English craftsmen before the end of the seventeenth century. From about 1715 repoussé work – ornamental metal work hammered into relief from the reverse side – became common as the usual form of decoration. Deeper repoussé work meant that a second plate was incorporated for the high relief work. There was a great variety of shape, style and decoration, although this was not entirely new. In 1575 Hans Kiening of Bavaria made a highly decorative book watch. Henry Sebut of Strasbourg signed, and so presumably made, a silver sphere engraved with a map of the world – but nothing else is known of Sebut. Another unknown, one Thomas Sande, made a tulip watch, Jacques Jaly produced a dog watch, and Johann Maurer made a fine skull watch; all three descriptions referring to the watches' shapes.

In Limoges in France the art of painting on enamel became popular in the fifteenth century and much use of it was made. The Hiraud family of Geneva were acknowledged masters of the craft. This technique was first class but the colours tended to be too harsh for some tastes. By about 1650 painting on enamel had gone out of fashion. Generally speaking, a watch had to be opened in order to tell the time, although crystal covers were not uncommon, but unfortunately they were insufficiently transparent. Then glass began to be used, but it is not known in what year.

WATCH MECHANISM

Early watches had verge escapements and the first improvement on it was the cylinder escapement. This was an idea of Tompion's but it did not appear as a practical proposition for another twenty-five years in the early 1700s. The cylinder escapement had little effect upon the English designs but became popular on the Continent. There it was soon possible to produce fairly inexpensive watches with escape wheels and cylinders in highly finished, hardened and tempered steel. The duplex was a frictional rest escapement which was invented in France about 1724 by J. Duterte. The idea was applied to many watches but it never became popular, probably because the duplex was difficult to make. However, many fine duplex watches were constructed and with them will always be associated the name of McCabe, an eminent and successful Irish watchmaker living in London in the early 1800s.

In Germany the stackfreed was employed even after 1650 but this mechanism was never used in France, people like la Garde preferring the fusee. The early fusee was connected to the spring-barrel by catgut but fusee chain was introduced shortly after 1600. Chains then became increasingly popular; by 1675 they were universal for watches and they continued to be used in bracket clocks until the nineteenth century. After 1575 the wheel balance in conjunction with a fusee became relatively common but, like the stackfreed, the dumbbell-shaped balance common in early German watches was still used until 1625. French and English watches generally had balance wheels and in earlier German watches they were carried in plain S-shaped cocks fixed to the top and bottom plates. From about 1575 the brass movement began to oust steel.

ENGLISH SUPREMACY

Improvements, of course, were on the way and the invention of the anchor escapement and then the balance spring gave English horology its supremacy from 1660 to about 1750. During this period England was fortunate to have a number of extremely able craftsmen. In many ways the greatest among them was Thomas Tompion (1638–1713). How few people are aware that although he is famous for his clocks, standardised though they were, he made only 650 of them. His output of watches was 5,500.

In the seventeeth century an interesting type of clock-watch appeared. About three to five inches in diameter and thick in proportion, these instruments were fitted with calendar and alarum mechanism. The earlier examples had a date aperture but on later models the date was indicated by a hand, the days of the month being marked between the hour and minute circles. Also in the earlier models the alarm was set by a disc in the middle of the dial with the hours marked I to XII.

Watch by Thomas Tompion of London, about 1675

Somewhere about the middle of the century England began to develop a lucrative export trade to Turkey. Few examples of the early exported clocks remain but after about 1700 the export of clocks and watches increased considerably. Not many of the leading makers made the clocks and watches for this particular trade. Lacquer or tortoiseshell cases with brass or silver mounts were popular in Turkey, as were hemispherical dome tops. The chapter rings carried Turkish numerals and in order not to offend the religious importers, spandrels showing angels and females' heads were left off. The importers were always interested in clocks containing small organs, musical and chiming movements.

DIALS AND HANDS

There was a considerable difference between English and French dials. The former used a standard pattern – the engraved dial – but the latter went in for a rather clumsy arrangement of separate enamel plaques for each hour, mounted on a metal dial plate. Before 1675 few watches were fitted with minute hands but the introduction of the balance spring made them popular. English hour hands were of a type known as beetle, being roughly shaped like a stag beetle; the minute hand was poker-shaped. The early hands were of steel and black in colour, but later on models were blue and usually more delicately made. Long after English watchmakers had accepted the idea of two hands, the French and Germans were still making single-handed watches. By the late 1820s the pocket watch had reached a high degree of elegance, but also the end of artistic clock-making in France had come.

Pierre Frederick Ingold (1787–1878), Swiss by birth, made some interesting machines for producing watch parts. He worked for some time with Breguet in Paris, then came to London to assist in starting the British Watch Company. Within two years it failed, probably because of opposition from the more orthodox clock- and watch-makers of the time. The nucleus of the firm went to America; and Ingold followed to New York in 1845. When the Americans and Swiss began to show a lead, valiant efforts were made in England to produce machine-made watches at competitive prices.

CLOCKS IN HOLLAND

One Dutchman, Huygens, made two important contributions to horology. However, his development of the pendulum and the balance spring seems to have failed to encourage his fellow Dutch craftsmen to take advantage of his work. The English subsequently took that advantage. It is interesting to note that while the Dutch did not produce a clock or watch towards the end of the seventeenth century which can be said to have a distinctive national style, they nevertheless had a strong influence at that time on English furniture and clockcase design. However, the Dutch did produce many bracket clocks, long case clocks and watches, and a number of their craftsmen emigrated to England.

What in England is called Dutch striking is called Double striking in Holland, and it uses two bells of differing pitch. The larger deeper bell denotes the hour and the stroke on the small higher-toned bell indicates the first quarter. At the half-hour, the next succeeding hour is struck on the small bell; at the third quarter one blow is sounded on the large bell. In what is called the Zaandam clock, the striking train does not operate the quarters, but cams are used – projecting parts of the wheel – which convert circular motion to variable motion, and which are fitted to an intermediate wheel geared to the minute wheel. Thus there is no warning when the clock is about to strike. These Zaandam clocks were first made about 1670 and were produced until the middle of the next century. Two angle brackets, supporting the clock, were fixed to a large wall case which itself contained the pendulum. These clocks were made with one or two hands, and ropes instead of chains were used for the weights.

There were also three other clocks with distinctive names. The Frieseland, from the early part of the eighteenth century, was a highly decorative but cheaper edition of the Zaandam clock, never signed by the maker and usually left undated. The Staartklok, a type of clock exported in large numbers, had its movement housed in a similar manner to the long case clock and striking was frequently done on two bells. The dial often incorporated a calendar and some astronomical observations. More finely made than the Staartklok but similar to it, was the Amsterdam clock, which was sold to the richer people.

FRANCE

French clock-making meant more than producing a movement which allowed the hands, turning against a fixed dial, to show the time. An integral part of the industry was the case and its finish. Clock-making included the cabinet maker, the gilders, the bell founders, the steelmakers and the bronze casters, the sculptor and the workers in marble, enamel and porcelain. Something had to be known also of fine metals and their composition. Not until the middle of the eighteenth century did the movement take its rightful place in the industry. This is surprising, for at times French-made scientific horological instruments were superior to those of every other country. French clock-makers were always in the forefront of design yet their weak-looking, simple movements attracted less interest than the complex and solid English counterparts.

When a suitable design was produced the French tended to keep it for some time and the small flimsy products of the Renaissance period lasted a century. Very few Renaissance clocks survive and this may be because the precious metal in the cases was needed for other purposes and the movements were destroyed. A favourite shape was the hexagonal and it lasted until the middle of the sixteenth century. From engravings of a portrait of Louis XI standing by a table that supports a small hexagonal type clock, it is believed that the clock carried the first coiled spring; the engravings also place the beginning of the French clock industry no later than 1483. These engraving are the earliest evidence from anywhere of the spring clock. Before the hexagonal shape gave way to a small square-shaped clock with a dome, in Henry IV's reign, an interesting development took place. This was the two-storey clock, each storey carrying a movement, the floor between the storeys being supported by minute Doric columns. A good example of this type of clock can be seen in the Richard Flagg Collection, Milwaukee. The cases usually included an entablature and dome with cast columns, pilasters or caryatids at the corners. The dial was mounted on the movement. This allowed the latter to be removed from the case easily without the dial being detached. Early clocks had movements of steel but towards the 1600s there was a change to brass, first for the plates and later for the wheels. It was probably German influence towards the end of the century which caused French clockmakers to break away from tradition and to produce a variety of designs together with a considerable improvement in the decoration, piercing and engraving.

By about 1650 the French industry had practically ceased. This may have been through the increasing interest in the pocket watch and the consequent drop in sales of the more cumbersome table clock. No one really knows. A great revival of the industry resulted later from Huygens's demonstration of his pendulum in Paris about 1657. The capital became the main centre though clock-making was also developed at Blois, Lyon and Rouen. The success of the pendulum gave the French clockmakers an opportunity to produce a new design in place of the Dutch styles so often copied. The new design lasted for some fifty years and the religieuse, as it is called, was and still is regarded as one of the finest-looking clocks ever produced in France.

Inspiration for design during Louis XV's reign had come from Rome but the designers during the following reign turned to Greece. The pendule d'officier, or travelling clock, was made in quantity during this period. The movement was enclosed in a square gilt case with small handles on top. The dial occupied the whole of the front of the clock. Great attention was paid to dial styling, and coloured dials – as opposed to plain white ones – now appeared. The size of the numbers was reduced and the minute numerals vanished. Arabic rather than Roman numerals for the hours gained in popularity.

A great variety of designs began to appear among the tall clocks. Presumably copied from the English tall clock, the French ones were never popular and few were made. However, there were plenty of rococo mantel clocks in bronze cases; there were asymmetrical mantel clocks surrounded by finely made figures of Meissen porcelain or flowers of Vicennes porcelain, sometimes both. Animal clocks became popular and among them were featured the horse, bull, elephant, rhinoceros, lion, boar and camel. During Louis XVI's reign (1774–89) clock cases – except for the few tall clocks – were no longer made from wood. Movements surrounded by metals, porcelain and enamel were greatly in demand. However, movements of the Louis XVI clocks began the golden age of French clock-making and it was the clockmaker, not the casemaker, who deserved the praise this time.

In France, as in England, the guild system was strong and influential and quality, quantity and production methods were jealously guarded. The earliest guild of French clockmakers dates from 1544 in Paris, and 1597 in Blois. As in England, the French guild system did much to improve the status of the craftsmen and the quality of their products. (An elementary form of mass production finally appeared in the eighteenth century; perhaps those early guild masters had foreseen the likelihood of some deterioration in the quality of later clocks and watches.) It is interesting to note that in Germany the reverse effects were the result of the foundation of the guilds. In short, guilds caused the downfall and near destruction of the German clock industry.

The early turret clockmakers had been locksmiths and they were too few in number to warrant the establishment of separate guilds. The most exclusive guild in France – and there were many guilds associated with clock-making and watchmaking – was the Paris guild, which limited the number of master craftsmen to seventy-two. Certain guild members enjoyed the patronage of the King and were thus beyond the authority of the guilds. They could open as large a shop as they wished, employ any number of apprentices and carry out work normally forbidden to other guild members. There were always three men in this privileged position and they were known as *Horlogers du Roi,* or Clockmakers to the King.

SWISS CLOCKS AND WATCHES

Although a number of countries now manufacture millions of clocks and watches, Switzerland is often thought of as the land of the best timekeepers. This is no longer true, but it is a fact that the Swiss clock and watch industry is highly organised and efficient, and produces more watches than any other country.

In the market square of Le Locle is a statue of a small boy

wearing a blacksmith's apron. An inscription tells us that this is 'Daniel Jean-Richard, founder of the watch industry in Switzerland'. As he had five sons and they all entered the new industry we can take it that the claim is well justified. Probably the reason for the well established industry is that in its earlier days Switzerland was predominantly a farming country and work was scarce during the long and hard winters. In the eighteenth century Jean-Richard and his family began what we would call a cottage industry. Whole families made different parts of the watches, which were then taken to a centre for assembly, regulating and marketing. By the middle of the nineteenth century, when Britain controlled the world market in clocks and watches, the Swiss introduced mass production. British watchmakers, centred on Clerkenwell, London, kept to the methods of the hand-made cottage industry and consequently soon lost their supremacy. At the end of the century the Swiss ousted the British from their superior position.

Today the Swiss watch industry, with a manufacturing figure of some forty-five million watches annually, still keeps in the main to the old system but the whole process of watchmaking has been organised to an extremely high degree. Specialist factories using complicated machinery produce the cases, others produce mainsprings, etc., and the Fédération Horlogère at Bienne controls Swiss watch prices, but in a way that prevents undercutting from affecting quality.

GERMANY

A seventeenth-century clockmaker of Augsberg, Johann Phillip Treffler, who made a clock for the Medici Palace, Florence, Italy, had, like Huygens, given thought to the pendulum as a time measurer. Little is known of Treffler but Mr S. A. Bedini of the Smithsonian Institution in Washington did some research and accorded the German the credit for the clock. Huygens, having obtained his patent from the States General in 1657, commissioned a clockmaker in The Hague, one Saloman Coster, to produce clocks to his specification, which

18th century north German oil clock. The glass reservoir is graduated from 8 pm to 7 am to show the hours of the night by the quantity of unburnt oil

naturally incorporated his new principles. The pendulums were hung with silken chords which were guided by brass cheeks. In Holland the clocks were known as Haagse Klokje; in France they were the religieuses already mentioned, larger and more highly decorated.

Germany developed a wall clock, spring driven and with a short pendulum swing in front of the dial. The latter was of embossed metal in the form of a dish (Telleruhr). A great centre of German clockmaking was the Black Forest where clocks were made by a system of divided labour. Individual workers specialised in particular parts of a clock. There were dialmakers and dialturners, wheelturners and chainmakers, gongmakers and brassfounders and most important of all – for they set the pace – the toolmakers.

The earliest German clocks were about 1640. Fitted with the foliot, copies are now being made on a commercial basis as a leading tourist attraction. But somewhere about 1740 the Black Forest clockmakers abandoned the foliot. Twenty years earlier they had incorporated a striking mechanism in their clocks and up to the end of the eighteenth century wooden wheels were still being fitted. The Black Forest wall clock was a great success and was soon being sold all over Europe; by the early nineteenth century it could be bought in Russia and the United States of America.

It was a relatively simple clock consisting of a top and bottom of beechwood, with four corner uprights also of wood and three vertical wooden plates with brass bushes to carry the pivots of the wheels. In its earliest form the clock had wooden wheels and as the number of teeth that can be cut on a wooden wheel is limited, each train would have had five arbors. This number was reduced to four on the introduction of brass wheels. However, the arbors themselves remained wood with wire at the ends to form pivots. The anchor escapement and long pendulum did not appear in these clocks until about 1750. The dials used were generally of painted wood, but during the next century it became the custom to use paper dials, which were hand coloured and then glazed over. The

Black Forest wall clock was eventually a most reliable timekeeper and in these days of plastics and shining metal there is something warm and homely about an all-wooden clock. Hasluck in the *Clock Jobber's Handbook* for 1887 said of the Black Forest clock that it was 'a most trustworthy timekeeper . . . No other country makes clocks of this type'.

Exports flourished and expanded considerably during the early part of the nineteenth century but this was to come to an abrupt ending when Chauncery Jerome, an American clock manufacturer, sent his first consignment of factory-made clocks to England. They were in rectangular cases about twenty-six inches by fifteen inches with a painted glass tablet in the door and weights enclosed entirely inside the case, and the clock itself could be hung or stood on a shelf. The Germans learnt quickly, for this competition began in 1842 and within a few years factory production was started at Schwenningen by Johannes Bürk. His clocks were mainly for the nightwatchmen of the towns. In Schramberg the Junghans family began the mass production of domestic clocks, and they knew all about American methods of mass production for one of the younger members of the family had worked for a time in an American clock factory.

THE CUCKOO CLOCK

This clock has probably fascinated more people than any other type of clock made. Yet it is quite a simple affair and the inventor could never have been aware of the great popularity his brain-child would receive.

About the year 1730 Anton Ketterer of Schönwald made his first cuckoo clock. His early clocks were the same as the usual Black Forest wall clocks and it is only since about 1870 that this type has taken on the form in which it is known today. Ketterer based his invention on the church organ pipe, using bellows to blow air through the pipe which held a reed.

MEETING OF THE NATIONS

In Britain, industrialisation, the establishment of free trade in the 1840s and the Great Exhibition of 1851 were partly responsible for the final switch from the craftsman's watch and clock to mass produced timepieces.

The lives of the workers, many of whom were now employed in 'manufactories', became controlled by the clock and the demand for cheap clocks was met when free trade permitted the import of thousands of clocks from France, Switzerland and the United States of America. The English clock- and watch-making industry suffered considerably. The Englishman, B. L. Vulliamy (1780–1854) often sent his watch cases to Switzerland where they were fitted with cheaper Swiss movements. Denison, another Englishman (later to become Lord Grimthorpe), complained, in the mid nineteenth century, that American and French clocks, which were better than their English counterparts but would not last as long, were destroying the long established English spring clock.

Parliament Square, London, showing the clock known as Big Ben, from the name originally given to its hour bell

In 1851 the Great Exhibition at the Crystal Palace, London, brought all the major clock-making countries together and horology was a section of Class X of exhibits. The novelty and variety of materials used seemed to be important although

Clock mechanism of Big Ben

many hand-made movements were on display. It is interesting to recall, remembering Lord Grimthorpe's remarks, that France, Switzerland and Britain shared almost equally most of the medals and honourable mentions awarded at the Exhibition.

BIG BEN

One of the most famous clocks in the world is known as Big Ben although this is not really the name of the clock at all.

A special clock was required to do justice to the new 316 feet high tower of the almost completed Palace of Westminster or, as it is generally called, the Houses of Parliament. Vulliamy, Queen Victoria's clockmaker, was invited to submit his ideas but did nothing about it. The contract was won by E. J. Dent, a marine chronometer maker. The final cost was about £22,000 (or about $52,000).

The hour bell, weighing just over thirteen and a half tons, was called Big Ben after Sir Benjamin Hall, the First Commissioner of Works. The clock, whose mechanism weighs some five tons, was set going in May 1859 and two months later became a striking clock. The thirteen foot long pendulum weighs six hundredweight and beats every two seconds. When the clock mechanism runs down the weights almost reach ground level. Above the four twenty-three feet diameter clock faces in a small compartment is the Ayrton light, named after another First Commissioner. It is lit to indicate that the Members of Parliament are in session.

4
Electric Clocks, Chronographs—and Beyond

With the aid of electricity we can, from a single source, control in various ways a number of timing devices which may be many miles apart. An electric current can be sent from a master clock every second and thus the pendulums of a whole range of clocks can be kept swinging in time with the pendulum of the master clock. By another method which has now been more commonly used, a pulse of current sent out by the master clock, say every half minute or minute, can energise the electromagnet of each distant impulse dial and so move the hands forwards a half or one minute.

It was during the second half of the eighteenth century when various inventors began to experiment with electricity as a means of power for various forms of signalling. There was little success by way of electrostatic effects and the first major breakthrough was the result of the work carried out by Volta (an Italian who invented the Voltaic cell in 1800) and Oersted of Denmark (the discoverer in 1820 of the magnetic effects of an electric current). In 1838–9 Wheatstone and Cooke introduced the first commercial electric telegraph in Great Britain. It worked by electromagnetism and their receiving instrument had five or six vertical magnetic needles mounted on a dial on which the letters of the alphabet were printed. A separate wire and coil that served as an electromagnet controlled each needle. A sending device sent an electric current through the wires, producing in the coils a magnetic field. As each letter was sent, the magnetic field caused a needle to point to that letter on the dial. This telegraph system was used until 1870. The American inventor and painter, Samuel F. B. Morse (1791–1872) had, like Wheatstone and Cooke, developed a successful electric telegraph. From this a number of scientists, including the three mentioned above, saw that it would be possible to transmit synchronising signals to a number of other clocks. In 1842 Hipp of Neuchâtel, Switzerland, showed that with the use of an electromagnet a pendulum could make an electrical contact and thus help itself to an impulse whenever the arc of swing fell below a given value. A constant swing is possible using this method whereas the strength of the alternative source of power, the battery which is not constant, will affect the impulses and therefore their frequency.

Two years before Hipp's exposition, Alexander Bain, a Scottish clockmaker, devised a number of ways of controlling distant clocks from one master clock. He did this by sending out pulses of electric current at pre-arranged times, eg every second or every minute. He also tried methods of keeping a pendulum in vibration by means of electromagnetic forces. His ideas were generally good but did not become commercial successes because the electrical contacts were not efficient enough. Bain also produced a method whereby the hands of a clock could be forcibly corrected hourly by an electric signal. Some thirty years later systems similar to Bain's were used by the Standard Time Company of London and Messrs Ritchie of Edinburgh. A system patented by Lund in 1876 was used on the railways and in more modern times a similar system has been used on the London Underground. Under this system the individual clocks have been electrically wound and corrected hourly, as required, by a time signal sent out from a controlled, central master clock.

There are four main types of electric clock.

1. Self-contained.
2. Master clocks containing a number of dials.
3. Electrically wound clocks.
4. Synchronous motor clocks.

A self-contained clock can be driven off a one-and-a-half-volt battery. The pendulum bob, which is a solenoid, swings

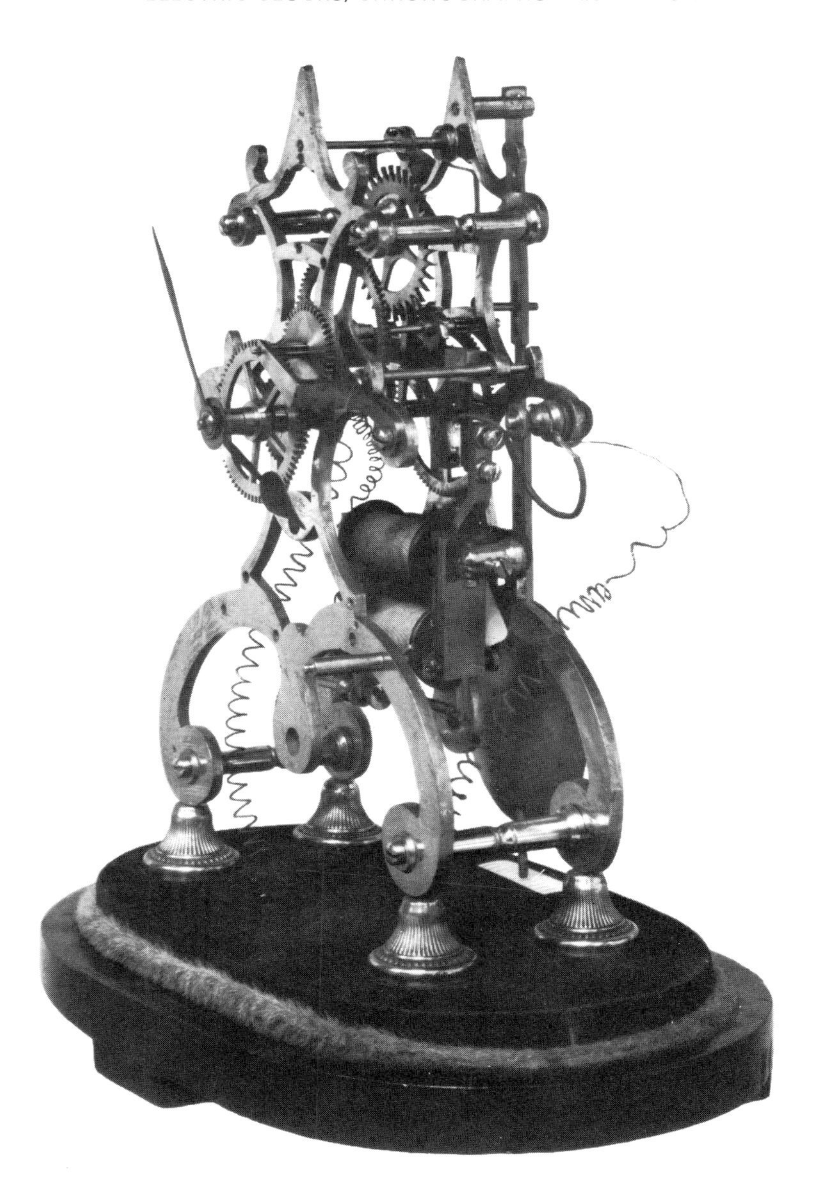

Self-winding clock of 1895: a mechanical clock driven by a weighted lever which is lifted every half minute by an electro-magnet, and employing a synchronome switch, a type now widely used in master clocks

over a permanent magnet and attached to the rod is a small silver pin. When the pendulum swings to the right the pin completes the circuit by coming into contact with a metal plate and the electromagnet gives the pendulum a kick further to the right. No return impulse is necessary.

The master clock system has two great advantages. First, all the clocks in the master clock circuit show the same time, and secondly, the system, when used in conjunction with a special type of pendulum called the Shortt Free Pendulum, can give a variation as little as one hundredth of a second a day. Such great accuracy has meant that the equipment can be widely used in many observatories.

Electrically wound clocks are simply clocks from which the main spring has been removed. This type of clock is commonly used in aircraft, ships, cars, street lighting etc. The pendulum is usually replaced by lever escapement.

The synchronous clock, which we use in our homes, is not a true clock at all for it is regulated by a master clock in a power station. Thus, it is only a repeating instrument, repeating the master clock's time. This master clock is set by the radio pips or Greenwich and the mains are an AC time-controlled electric current. If the generators are running at their normal frequency the hands on the clocks fed by the power station will progress at the proper rate. The disadvantage is obvious – power reduction or a power cut renders such clocks useless. It is possible, however, to fit self-starting clocks with an indicator which will show if any such interruption has occurred since its last setting.

Some fifty years after Bain's experiments, it was suggested that alternating time-controlled electric current could be used to operate clocks. An American, H. E. Warren, put this idea into practice and it was not long before many houses in both America and Britain were wired for the clock point.

THE SHORTT FREE PENDULUM

About 1921 the Shortt Free Pendulum first appeared, installed in Edinburgh Observatory, and under good conditions it had a daily variation rate of only a few thousandths of a second; a variation of less than one part in ten million. It was W. H. Shortt in conjunction with F. Hope-Jones and the Synchronome Company who produced the Shortt clock and it was not superseded for many years until the quartz crystal clock appeared.

The free pendulum does not have to turn the count wheel as this task is performed by a 'slave' clock. Every thirty seconds the slave clock releases a light arm carrying a jewel which then falls upon a small wheel mounted on the free pendulum. As it rolls off it imparts a light impulse to the free pendulum and afterwards transmits a synchronising signal to the slave clock by means of Shortt's 'hit-an-miss' synchroniser. The free pendulum swings entirely free except for the fraction of a second every thirty seconds when receiving its impulse. The impulse is constant in amount, of course, and is given symmetrically about the midpoint of the swing.

THE QUARTZ CRYSTAL CLOCK

The first quartz crystal clock was made in 1929 by Horton and Marrison of the United States of America. It was designed originally to provide a smaller unit of time than the astronomical clock and is capable of measuring frequencies which are met with in radio engineering. Now it is gradually taking over the function of the pendulum. It has been found possible to link the quartz clock with the natural vibrations of atoms and molecules, for the radio frequencies emitted by atoms and molecules at low pressures are fixed and unchanging with time. Therefore, such frequencies can be used to control an oscillator which itself controls a quartz clock.

L. Essen and J. V. L. Parry at the National Physical Laboratory in Teddington, near London, used the vibrations of the caesium atom to control the quartz clocks at their Laboratory and at the Post Office standard frequency broadcasts. This was the first clock to give a practical improvement over the quartz clock and a more recent version of the atomic clock is accurate to the equivalent of one second in 300 years.

One of the uses of the atomic clock is in measuring the rate of the rotation of the earth, which, it has already shown, may vary by four or five milliseconds from one day to the next. The unit of time based on the daily (diurnal) rotation of the earth (known as the mean solar second or second of Universal Time) is no longer accurate enough for some modern scientific and technical work. In 1956 the International Committee of Weights and Measures adopted a new unit known as the Ephemeris Time. This is based on the tropical year, the time interval between successive returns of the sun to the particular point on the celestial equator that is known as the vernal equinox.

A fully electronic clock has been produced by Solvis & Titus S.A. of Geneva. This clock, called the Soltronic, has no moving parts at all and the time is indicated by lights. The mains electricity, battery or other source provides the power and the moving electrons inside the clock do the work of the mechanical parts of a conventional clock. It is possible that electrons can be made to work the whole range of clocks from city hall clock to elegant wrist watch. It is already possible to have a clock without the ticking which some people find irritating. The Horstmann Gear Co, Bath, England, have manufactured a silent clock which yet keeps pendulum control by using a magnetic escapement. By using magnetic attraction the pendulum controls the rate at which the train runs.

SUPER WATCHES

In an attempt to produce an electric watch, Lip Besançon of France and the Elgin Watch Co of the United States have collaborated. However, the first successful electric watch, powered by a battery, was made in 1957 by another American company, the Hamilton Watch Company.

An ingenious idea was tried by Max Hetzel, a Swiss electronics engineer. He used a vibrating tuning fork. At that time he was employed by the Bulova Watch Company, USA, and there was able to produce the Accutron. This watch, with about half the moving parts of a normal watch, uses an inch-long tuning fork in place of the usual escapement, and a 1·3 mercury cell to power the transistorised oscillating circuit – a highly ingenious device. Yet it is Hattori, Seiko-Hattori of Japan who seems most likely to succeed in producing the first atomic wrist watch. This watch is powered by a cylindrical battery half an inch in diameter and a quarter of an inch deep. The battery was developed to Hattori's specifications by the McDonnell Douglass astronautics company of America and it works on energy released by a radio isotope of Promethium 147. With a life span of five years and a replacement costing about $4.50 this watch looks like having a promising future. Another obvious advantage over conventional self-winding watches is that it has fewer moving parts, and that means cheaper production and less mechanism to go wrong. Beta rays produced by the battery are prevented from leaking by a housing of tantalum and a stainless steel cover. Safety regulations both in manufacture and in the end product are stringent yet such a revolutionary watch is, initially, bound to have its opposers.

There is no doubt that with miniaturisation of electronic equipment, batteries etc, the days of the mainspring and escapement are numbered. Clocks and watches may take on new shapes and styling and be produced in a variety of materials, but with clocks at least, such old and beautiful shapes as the lantern clock will, we hope, remain with us. There is plenty of evidence that it may, and why not? A lantern clock – with modern works – is a pleasing shape to look at; a swinging pendulum more homely than a length of flex nailed to the wall.

AUTOMATIC WINDING

The ideal watch, of course, would be one that required no winding up. We are lazy creatures. A number of watchmakers have given their attention to this problem and the obvious solution would be to create a situation in which the mainspring can be rewound by the wearer's arm movements. Recordon, a watchmaker and Brequet's London agent, patented an automatic winding watch as far back as 1780 and he made a number of such watches. Little more was heard of this type of watch until 1924, when John Harwood, an English watchmaker, succeeded in making an automatic winding wristwatch and took out a Swiss patent. In such a watch as the automatic winding type the force of the mainspring tends to be more or less constant while the watch is being worn. This means greater accuracy but it can now be achieved in other ways.

TIME SIGNALS

The British Post Office Telephone Department's TIM and the six pips ensure accurate time for the caller at any hour of the day or night. This system of time broadcasting arose out of a suggestion, made about one hundred and ten years ago, that Greenwich could be used for giving the whole country the correct time. Greenwich was, in fact, relaying time to the Post Office and great progress has been made since; time is now available on 'tap' to everybody.

In 1924 a radio announcer used to give the time at 10 pm each evening. He counted 5 . . . 4 . . . 3 . . . 2 . . . 1 from his watch but the sponsor of the idea considered this to be inaccurate and the following system was evolved. In the control rooms, BBC engineers have clocks regulated by a master clock which is itself checked every fifteen minutes by Greenwich. At a minute to the hour or quarter hour on which the signal is to be broadcast electrical contact is made with Greenwich, and at eleven seconds to the time the regulator clock at Greenwich does the rest. The 'pips' are the oscillating of a valve at a set frequency and the interval between each is one second, the last of the six pips giving the hour. It should not be confused with Big Ben where the first stroke denotes the exact time. Originally control was by a Dent regulator clock but this was successfully substituted by phonic motors in 1949 and by entirely electronic means in 1957.

Twelve years after the introduction of the pips, TIM became available to the public. This was the result of a visit of a number of Post Office Engineers to Paris where the 'Speaking Clock' had recently been introduced. It was realised that this could be a valuable service to the general public. Behind the simple operation of dialling TIM there lies a vast network of electrical apparatus. There are two pairs of clocks housed in London and Liverpool and telephone calls are connected manually or automatically to the appropriate clock. By the use of short phrases recorded by an operator on a number of discs, and the interchange of these discs, it is possible to get a variety of announcements. By complicated radio circuits, which include photocells, amplifiers and warning systems, the exact time is given at ten second intervals throughout the twenty-four hours. The process is an automatic one and it is not a live voice which tells you the time when you dial TIM.

SUMMER TIME

It is taken for granted in Britain that during summer months the evenings are longer partly because the clocks are put on an hour and partly through natural causes. But our enjoyment of the longer evenings is due to the tireless efforts of one man.

A great deal of thought was given to the idea of making the most of daylight hours by William Willett, a London builder, and in 1907 he published a pamphlet entitled *The Waste of Daylight*. He showed the great many advantages of putting the clocks on during the summer time when there is more daylight than during the winter months. He used every argument to try to convince Parliament, local councils and the general public and there was much to be gained by inaugurating his scheme. The advantages to health and happiness were obvious, Willett pointed out, and he showed that something like two and a half million pounds could be saved by having an additional 210 hours of daylight a year. There was nothing revolutionary about his ideas, he explained, for standard time had been altered in Australia and South Africa in the 1890s.

Among the supporters of the scheme was Sir Winston Churchill, and he addressed a large meeting at Guildhall on the subject. The watch and clock industry was divided over the question. It was said, quite wrongly, that it was a bad thing to put a clock forward or backward, particularly if it was a striking or chiming clock. On the other hand electric clock-makers favoured the Daylight Saving Scheme, for as far as they were concerned, it was a simple matter to adjust all electric clocks from a master clock.

It is strange that some of the greatest opposition to the scheme should come from the agriculturalists, but Willet was a man of great determination and finally, in the year after his death, his efforts were successful. As a wartime economy measure a Daylight Saving Bill was passed in 1916, the Bill becoming law in 1925. During World War II clocks were advanced two hours instead of one, and ordinary summer time, as opposed to Double Summer Time, was retained throughout the winter. The dates when Summer Time begins and ends are altered to suit requirements. For example when there was a fuel crisis in 1947, Summer Time started on 16 March and ended on 2 November, and Double Summer Time was from 13 April to 10 August. Thus less fuel was used for lighting as there were more hours of daylight. Many countries have now adopted Summer Time.

In order that Britain should fall more into line with Continental business hours, a new experiment was tried out from 27 October 1968. Under this scheme Britain lost an hour throughout the day by advancing the clock one hour throughout the year, which meant that children went to school in the dark and farmers began work before daylight. There were advantages, for a Paris businessman, wishing to start work early, could telephone his London counterpart and find him in his office, whereas under the old system he would have had to wait at least an hour for London time, as it were, 'to catch up with Paris'. However, in 1970 the British persuaded their Members of Parliament to reject such Continental 'gimmicks' in a free vote in the House and to return to the saner scheme of 'daylight at breakfast time'.

TWO DAYS AT ONCE

Because of the rotation of the earth local time varies from

place to place, being earlier if we travel eastwards and later if we go westwards in relation to the time at the Greenwich meridian. In other words time changes according to longitude and one hour corresponds to about 15° shift in longitude. In

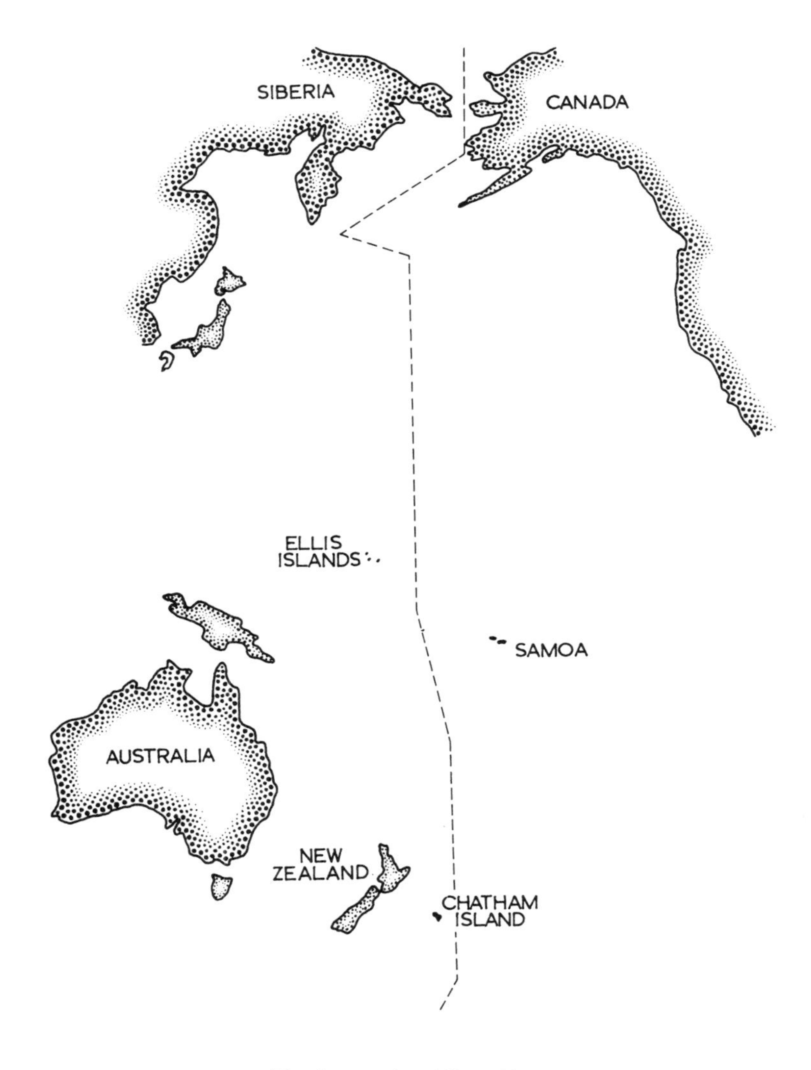

The International Date Line

a country as small as Britain this makes little difference although even Bristol is ten minutes behind Greenwich time, and most small countries use a universal standard time throughout.

Meridians are the lines of longitude and in order to standardise measurements of longitude it was found necessary to reach international agreement on which particular meridian should be used from which measurements could start. So in 1884 geographers met at Washington D.C. in the United States and agreed that the prime meridian should be that meridian which passes through what was then the Greenwich Observatory, England. In 1958 the Royal Observatory completed its move from Greenwich, where it had been since 1675, to Herstmonceux Castle in Sussex. The choice of England was not really so unusual for at that time Britain was a seapower of enormous authority and the Thames a river of great prosperity. So, from this line – marked 0° on the atlas – longitude is measured in degrees, minutes and seconds along a line of latitude, an imaginary line running west-east. When the sun shines directly down on a meridian the time is noon all along that meridian. The word meridian comes from the Latin *meredies*, which means midday.

Although the idea of Greenwich being the zero or prime meridian was universally adopted, it is not convenient to have Greenwich Mean Time on all clocks in all lands. So the larger countries were divided into time zones and their clocks based upon Greenwich Mean Time. If, for example, you travelled eastwards from Greenwich you would finally arrive at a spot which was apparently 12 hours earlier, and a yard or so further on you would have lost a day. This line, longitude 180°, is where we are 12 hours ahead or 12 hours behind time.

This, the 180° meridian, which is exactly half way around the world from Greenwich, marks the place where each new calendar day begins. The International Date Line follows this meridian for most of its distance. Here and there it zig-zags in order to avoid having two calendar dates on the same day in any country. The sun appears to travel over 15° of the earth's surface during each hour. Places to the west of the International Date Line or 180° meridian are behind Greenwich Time and those to the east of the International Date Line are ahead. Thus Australia is nine and a half hours to ten hours ahead of Greenwich, and New York five hours behind; while if it is noon on Saturday in London it is midnight on Friday or Saturday on the International Date Line. The direction in which the traveller is travelling will, of course, determine which day.

So there is a twenty-four hour time difference between the two sides of the 180° meridian. If a ship crosses the line in a westerly direction it loses a day; if it crosses in an easterly direction it gains a day. To determine a local time it is necessary to know the exact longitude and the time at Greenwich. Then divide the longitudinal degrees, minutes and seconds by fifteen (because one hour of time corresponds to fifteen degrees of longitude), and the answer is the time in hours, minutes and seconds ahead or behind Greenwich. For example, Malta is almost one hour ahead of Greenwich for its longitude is approximately 15°E.

When it is noon at Greenwich it is:
1 pm in Berlin, Vienna, Tunis and Nigeria
2 pm in South Africa and Turkey
3 pm in Aden, USSR and Iran
4 pm in Mauritius
6.30 pm in Burma
7 pm in Indo-China
8 pm in Hong Kong and Western Australia
9 pm in Korea and Japan
10 pm in Queensland, Victoria and Australia
Midnight on the International Date Line
1 am in Samoa
2 am in Alaska
2 am in Yukon
4 am in Vancouver
5 am in Salt Lake City
6 am in Winnipeg and Mexico

7 am in Montreal, New York, Washington and Panama
8 am in Trinidad, Chile and Argentina
9 am in Brazil
10 am in the Azores
11 am in Iceland and Senegal
Noon in Belgium, France, Portugal and Morocco.

Most major countries have adopted the plan of reckoning time from the Greenwich meridian. In large countries such as Canada and the United States where local times differ, because of the great distances between east and west, it is not practicable to employ a single standard time throughout the country so it has been divided into zones. For example, Canada has five time belts or zones.

TIME RECORDERS AND SWITCHES

A time recorder is an instrument for recording automatically the time of day at which a certain event takes place. The earliest of such devices was probably the watchman's clocks which were invented by Whitehurst of Derby, England, about 1750. Recording was simple. A wheel rotating with the hour hand carried a number of pins which passed one by one beneath a striker. If the striker was hit it forced the pin below it inwards and by looking at the position of the pins it was possible to see at what time an event took place. A recorder which could print a record in figures was invented by Bundy in the United States of America in 1885. This machine was superseded within some three years by a dial time register invented by Dey of Aberdeen. The most common type of recorder is the punched card type used by employees for 'clocking in' at work. This was an American development by Cooper in 1894.

The time switch is something we see working in the streets and shops of our towns and cities. Briefly, it is a switch which is automatically operated by a clock. Messrs Horstmann produced a time switch in 1914 which could switch lights on and off automatically as the times of sunrise and sunset varied according to the seasons. Time switches turn the street and shop lights on and off, they control the lights on buoys, the factory hooters etc. We always associate the time switch with electricity yet there are on record two instances where time switches were used to control gas street lamps. Dr Thurgar's invention of 1867 may not have been very successful but thirty years later Bournemouth gas street lamps were controlled by a switch invented by Mr Gunning of that town.

Since the early 1950s there has been an even greater interest in athletic and other sporting events throughout the world. This has increased the interest in record breaking and the consequent demand for more and more specialised time measuring equipment of ever great accuracy. Seiko Watch-K. Hattori & Co Ltd of Tokyo, Japan, has studied this problem. As a result of research the company was designated 'Official Timer' for the 18th Olympic Games held in Tokyo in 1964, and has so performed for twenty-four other international sporting events including such meetings as the 1965 World Wrestling Championships in Manchester, the 1968 Macao Grand Prix, the 6th Asian Games held at Bangkok in 1970, and the Winter Olympic Games in 1972 at Sapporo. Seiko has produced a complete line of electronic and electric sports timers and systems for use in a very wide range of events under all kinds of weather conditions.

For the 1970 Japan World Exposition, an entirely new and revolutionary timing system was devised which utilised atomic frequencies with an accuracy equivalent to plus or minus one second over 1,000 years. The photograph on page 86 shows the digital electronic timer used in the Asian Games. Operated by a built-in long duration crystal oscillator designed to be used as a time standard, the instrument is accurate in temperatures ranging from 50°C to –20°C. Conveniently placed push buttons minimise operational errors. Other features to prevent mistakes include a red 'in operation' light as well as a window for digital numbers, which are shown in red by a diode to ensure easy reading. The red light diode makes the timer a semi-permanent operating unit, powered by re-chargeable nickel-cadmium batteries. Resolution capacity of this instrument is one hundredth of a second. In numerous sporting events photo-finish equipment is essential and the photograph shows that time can be recorded to one hundredth of a second.

Digital electronic timer by Seiko-Hattori of Japan

Man has already shown that he is able to launch himself into space though his methods of so doing are expensive and clumsy compared with the sophisticated atomic and electronic equipment yet to come. Nevertheless, long before the first man left earth both the United States of America and the Soviet Union had to produce timing equipment which needed to be of the greatest accuracy. Hattori of Japan cooperated with

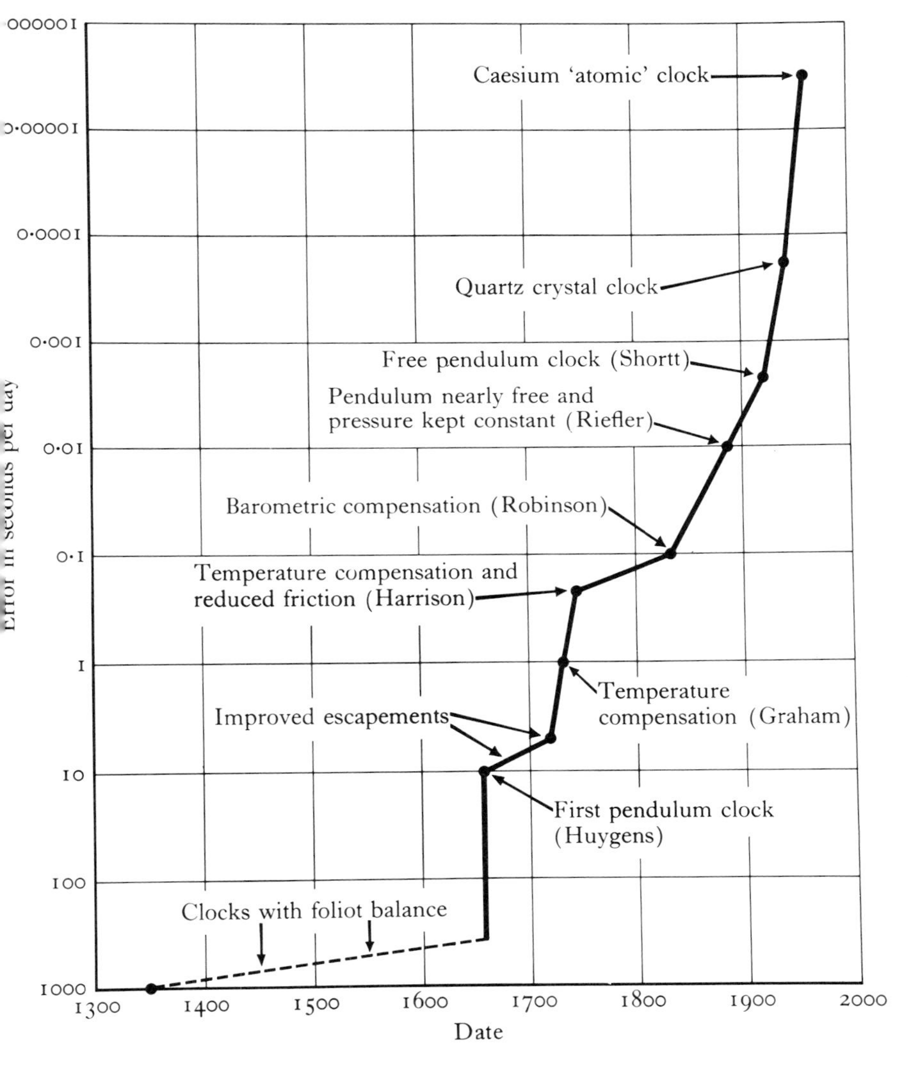

Chart showing progress in the accuracy of timekeepers through the centuries

McDonnell Douglas and have produced an atomic watch; France and America collaborated to produce an electric watch; and Markowitz at the U.S. Naval Observatory, Washington, and L. Essen at the National Physical Laboratory, Teddington, Middlesex, arranged a joint programme for determining the relationship between the atomic and astronomical units of time.

International cooperation may yet help us to grapple with deep space, greater speeds and an understanding of light years.

CHRONOGRAPHS

If we wish to measure intervals of time of the order of fractions of a second, then an ordinary watch will not do. For this purpose we use special types of instruments called chronographs. With them it is possible to time races and the speed of missiles, to measure depth and to record the passage of the stars across the heavens. More and more use is being made of chronographs in psychological and physiological research into our reactions to any given situation. The simplest form of chronograph, and the one which most of us are familiar with, is the stop watch, and such watches were first made in the eighteenth century. By the pressing of a button the train of clockwork may be stopped; a later development was the addition of a mechanism which, when a button was pressed, permitted the hand to stop, start and fly back to zero. As these pocket chronographs, as they were called, are operated by hand then there must be inaccuracies which may well be of a second or two. So, in order to measure the time taken to run a race within a tenth or a fiftieth of a second, or to complete a given number of laps of a motor racing circuit, it must be possible to do three things: 1. record the beginning of the race; 2. record the end of the race; and 3. link these facts together automatically.

In 1840 the British scientist Charles Wheatstone set out to solve this great problem (and two years later worked on it again with South and Purday). A stop watch button was operated by an electromagnet, controlled by an electrical signal transmitted from the event to be recorded. The 1842 experiments carried out at Campden Hill Observatory measured the speed of a bullet fired from a pistol and the rate of fall of a ball under gravity. Wheatstone was clearly satisfied with his findings for he claimed that his electromagnet instrument was accurate to one sixtieth of a second. Furthermore, he claimed that the instrument could be used for measuring the velocity of sound in air, water or rock masses. At about the same time the grandson of the famous A. L. Breguet of France, L. F. C. Breguet, also claimed credit for having invented the electromagnetic chronoscope. In 1845, three years after Wheatstone's experiments, Breguet published an account of his own instrument, which he said he had made in 1844. There is of course nothing unusual in this, for many scientists, unbeknown to each other, have been carrying out almost identical experiments of one sort or another at the same time.

A few years later Hipp of Switzerland improved on all previous chronoscopes and claimed for his instrument an accuracy of one thousandth of a second. Like Wheatstone, Hipp used clockwork in continuous motion but he introduced a number of refinements which gave him the accuracy that he claimed. It had been suggested by Thomas Young as early as 1807 that a rotating arm could be used for measuring short intervals of time. He proposed to drive the drum by means of a suspended weight, the weight of revolution being controlled by a type of centrifugal governor. Now if the drum was covered with wax or paper upon which a pencil or strip of metal was gently pressed and could be deflected by hand at the required intervals, a time scale could be marked out on the record by means of a style fitted to a vibrating body.

Siemens in Germany had been well to the fore in the chronographic world and after a period of time produced a drum chronograph in which the recording was carried out by an electric spark. Probably the most successful chronograph for the artillery-minded of the day who were interested only in the velocity of projectiles was produced by the English clergyman F. Bashforth in 1865. He allowed the drum to revolve freely while measurements were being recorded.

In 1849 Washington Observatory first used the chronograph to record the transit of stars. These records were made on moving strips of tapes as in Morse telegraphy, or on a rotating drum, and marks were made every second by means of an electrical contact system on an accurate clock. This same method of recording was later introduced at Greenwich years later by Airy, who in 1856 installed a large chronograph using a conical pendulum to drive the drum uniformly. Before this time astronomers had timed their transits by listening to the beat of a clock, then estimating by ear the fractional parts of a second. Rather a hit and miss effort, it would seem.

It was soon realised that to obtain an accurate time scale either the motion of the recording surface must be kept extremely uniform or a fine time scale must be marked on the record.

A variety of chronographs then appeared, among them the moving-coil galvanometer by Kelvin, a Belfast scientist, in 1867. An ingenious device was produced by Blondel in France in 1893 and by Duddell in England in 1897. Not so long afterwards it was obvious to the scientists that loudspeaker units and telephone receivers could be adapted for use by fitting them with a writing style or recording mirror. A final stage was the use of the cathode ray tube, where a beam of light can be made to traverse the full width of a screen in one ten-thousandth of a second, or less if so required.

As we have seen, the Japanese have taken time recording to even greater degrees of accuracy for everyday purposes.

BRITAIN'S CLOCK INDUSTRY TODAY

This brief history of the measurement of time would not be complete without a reference to modern methods of watch and clock making in Britain. In the last chapter we shall see something of the American and Japanese stories.

War, always the succour of industry, came in 1914 and 1939 and from the stimulus it brought, the British clock industry revived. At the end of the nineteenth century it was experiencing most serious competition from Swiss and German manufacturers, who were already using mass production methods.

The superior craftsmanship of the British clockmakers, handed down by men like Tompion, Knibb and George Graham, could not compete with cheaper continental clocks and watches. The two wars, however, forced Britain to use mass production methods and time has proved that little has been lost by so doing. The following example illustrates the resourcefulness and ability of the English horologist.

Less than a century ago Samuel Smith made watches by hand, and his production could be numbered in scores only. Honoured by four Royal Warrants, and selling from a small shop in the Strand, London, the firm of S. Smith and Sons Limited were – less than fifty years later – making high class chronometers for the Admiralty and for foreign governments. They also produced a popular guinea watch. From such humble beginnings the firm became renowned. Scott and Shackleton used their products on their famous Arctic explorations, and the present firm of Smiths English Clocks Limited have seen one of their watches taken to the top of Mt Everest with Sir Edmund Hillary; others fly faster than sound and others return unharmed from far below the surface of the sea.

The demand for accurate time-keeping equipment is insatiable, and by the use of mass production methods one factory alone, belonging to Messrs Smiths, is capable of turning out an alarm clock every two and a half seconds. Samuel Smith, Thomas Tompion and the other great horologists would be amazed and indeed fascinated if they could visit a modern watch and clock factory. Their conducted tour would surely make them breathless. In the designing and drawing office they would see hundreds of drawings, fifty or a hundred times larger than the final watch or clock: drawings and cross sections of every minute piece that is used. There would also be models of the new designs, a research chemist to test metals, jewels and oils, designs for new tools with which to make the watches and clocks, and finally, the assembly lines. It is only by such accurate production methods under strict hygienic rules that the modern manufacturer can market clocks and watches which can hold their own with the precise handmade timekeepers of bygone ages. Much money and thought are expended upon new models, for it is more profitable to sell an

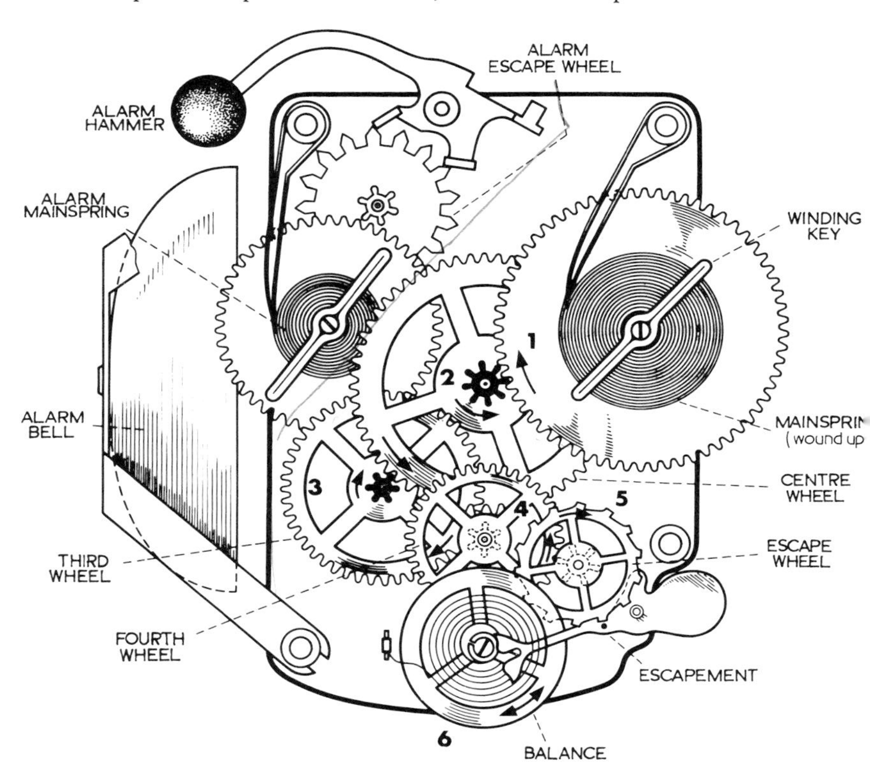

The movement of a modern Smith's alarm clock

accurate clock or watch than it is to have it returned to the makers because of faulty workmanship.

5
America and Japan

The settlers from Europe spread along the eastern seaboard of North America and built their townships as they went. Town records mention public or tower clocks as early as 1650 and although we unfortunately know little of their history we may be quite certain that one of the first things a new township wanted was a clock; a public clock. This was a considerable advance on the noonday cannon or the ringing of a bell. In the seventeenth century, the golden age of clockmaking in England, the pendulum with anchor or deadbeat escapement had replaced the less accurate foliot balance. Then the clock was able to give far more accurate time.

The early settlers also took with them their personal clocks, if they were lucky enough to own one, and as many of the new arrivals were from England lantern clocks and eight-day long-case clocks were the commonest type. So the new Americans had their town clocks and their private clocks. But things were very different in the new land. Among the settlers were locksmiths and other craftsmen who began looking forward to the time when every town would have more than one clock and every householder would have at least one clock.

Metals were scarce so the clockmaker advertised for scrap. Glass was in short supply and ways of producing it had to be devised. In fact, everything had to be made by hand during a period of makeshift. The craftsman, perhaps assisted by an apprentice, worked with the simplest of tools – the hammer, drill and file. Clockmaking was a tedious, painstaking, wearing job. Brass plates and blanks were needed and these had to be cast in moulds. Many craftsmen among the new settlers had their own formulae for the making of brass. It is an alloy of zinc, copper and sometimes tin, and after casting it had to cool but it would still be too soft for practical purposes. The only way to harden brass is to subject it to a long period of planishing, which is giving the brass light blows with a hammer. This method might take days and then cracks could appear in the brass. Some manuals of instruction on the production of brass recommended that when a crack appeared the planishing should be continued under water. Hardly a pleasant wintertime occupation.

Filing, polishing and smoothing were then necessary. Much of the hard work was taken out of the filing by the introduction of a process for casting brass between two polished slabs of marble which gave the brass a more uniform thickness. Later still, of course, the clockmakers could buy the brass castings and blanks ready for filing, cutting and finishing. A most exacting job was gear cutting, and even this was simplified when Elisha Purington, a clockmaker from Kensington, New Hampshire, devised a lathe-like machine for cutting teeth on the gear blanks. Some of the old clocks still bear the marks of the lines used for cutting teeth by hand before Purington's invention.

Thus a clock took many months to make and few people could afford one. So clock-making tended to be a part-time occupation and clocks were made to order. Nevertheless during the eighteenth century tall clocks were made in small but steadily increasing numbers. This was most noticeable in Philadelphia and the surrounding area. In spite of the primitive equipment used, finely made brass movements were set in equally splendid examples of the cabinet and case maker's art.

Clockmaking ceased when the War of the Revolution broke out and warlike materials were produced instead. So far little has been written of the part played in that war by the clockmakers and blacksmiths turned gunsmiths. When peace came again clockmaking was resumed but shortages coupled with increased demand forced up the price of the tall clocks. So a number of widely scattered clockmakers looked around for a cheaper and smaller clock. This created a problem, for up to

Banjo clock by Aaron Willard of Massachusetts, USA, late 18th century

that time American clockmakers had closely followed English styles. One of the earliest departures from the traditional type of English clock is to be found in the 'Case on Case' clock, which appeared about 1780 (at about the time when so many French and German names appeared among the 'new' American clockmakers). There seems to have been an attempt to make a substitute for the spring driven clock of the period without the complication of springs and fusees. In fact, although the Case on Case clock looked like any other spring driven clock standing neatly upon its own cabinet, the two parts constituted a single unit. We do not know where this particular type of clock originated but it may well have come from France. In that country it was the fashion to mount a clock on a pedestal so that pedestal and clock appeared to be two separate units, though, of course, the pedestal served as a cover or screen for the pendulum. The New England craftsmen made a complete break and developed native American types and styles. The Willards in Grafton, Massachusetts, produced a number of types, their finest being perhaps the Banjo clock, which then sold for some thirty-five dollars. Metal was still scarce and movements took considerable time to make yet Simon Willard managed to make some four thousand clocks between 1802 and 1840.

In Connecticut other clockmakers were thinking about the cheaper type of clock, but they foresaw that some form of mass production would make it possible for every home to have a clock. These clockmakers reasoned that as metal was scarce and expensive but wood was not, the solution was to produce a wooden clock. Gideon Roberts was soon in business with his all wooden Wag-on-Wall clock. At first it cost some twenty dollars and the tall case another twenty, but he was able to reduce prices later on. Roberts had lived in Wyoming Valley, Pennsylvania, at one time, and it is possible that he learnt from German and Dutch settlers the art of constructing clock movements entirely from wood. He appears to have been the first clockmaker of Bristol, Connecticut, to use simple mass production methods in his clock production. His place in the history of American clockmaking is assured for he also owned an assembly plant in Richmond, Virginia, which, after 1800, was operated by one of his sons. Undoubtedly Roberts did influence the future of the industry but to what extent only research will show. About the same time as Gideon Roberts was producing his Wag-on-Wall clocks, John Rich of Bristol, the Merrimans, the Ives and certainly Levi Lewis were all making similar clocks.

However, Eli Terry now appeared upon the American clockmaking scene. He can be given the credit for being the first man to mass-produce low-priced clocks – but at the same time he destroyed for ever individual craftsmanship. In 1806 he accepted an order for 4,000 clocks, then had to devise a sales technique which would ensure their distribution. He was successful and so began the factory system of production. By this time also, water power was replacing muscle as a driving force for machines and this changeover naturally helped Terry. Penrose Hooper said of Eli Terry, 'He was the last of the craftsmen and the first of the industrialists.' The Terry clock dial had the minutes marked inside the hour figures, which was a relic of the seventeenth century. The hands were usually of the same length but the hour hand carried a longer design.

Eli Terry had ideas and he was in the clock business to stay. He used aggressive selling methods which assisted in his successful sale of American clocks. To some extent he was also helped by the Jefferson Embargo of 1807–9 and War of 1812–14, both of which severely restricted the importation of material from English factories. In 1814 he produced a twenty inch tall shelf clock which he patented in 1816. This was followed by the beautiful Pillar and Scroll clock, which ran thirty hours on one winding. It sold for about fifteen dollars at the time and Seth Thomas purchased the rights to make it. A favourite was the eight day shelf clock with wooden or cast brass movements, with the weight lines carried over pulleys to give the maximum fall. This was a popular feature in later American clocks. The lower part of Terry's clocks was fitted with a glass plate, on which was painted some kind of design but on which was left enough clear glass for the swinging pendulum to be seen. Cast brass eight day shelf clocks were also made in quantity.

There was a boom in clockmaking in Connecticut during the 1820s and soon clocks from that district were being sent all over the country. Then a peculiar thing happened in American economics – clocks were used for barter in place of money. Indeed, it was possible to buy and sell houses for an accepted number of clocks. This crazy situation stopped abruptly when the depression of 1837 brought the American clockmaking industry to a complete standstill. It also saw the end of the wooden clock movement.

In an attempt to revive the industry – which had already shown that it was capable of making its own individual contribution to horology – Chauncey Jerome's brother, Noble, introduced a thirty-hour, weight-driven clock movement made from rolled brass strips or stampings, and enclosed the whole in what we call an OG case. (OG, or ogee, is a moulding found in both architecture and furniture, showing in section a double continuous curve, concave below passing into convex above.) These originally styled clocks sold at a low price. A million a year were sold and Connecticut-made clocks captured and revived the American market. As a result, the tall clocks of Pennsylvania, made between 1840 and 60, disappeared. The twenty years from 1837 were years of great prosperity for the clockmakers of Connecticut; factories and mass production were now well established. In Connecticut the clockmakers began using the coiled spring for motive power, and they were aware that they were no longer restricted to the shape of the cases they had previously designed for their movements. The sober lines which had been dictated by the weight-driven movements gave way to fanciful shapes. It was from Connecticut again that clocks controlled by a balance wheel – called the marine movement – appeared upon the market. The principle of this oscillating balance wheel had been developed independently by Charles Kirke, Silas Burnham, Bambridge, Barnes and others. These clocks had the added advantage that whereas the pendulum-controlled clock was a fixture, the balance wheel type could be moved from place to place. This was an enormous step forward. The next stage in clock construction, which took place somewhere around 1875, was the round, metal-cased lever-escape alarm clock. This was the ancestor of the modern alarm clock and made up much of the American clock production figures for some years to come.

Up to about 1845 American clocks, like the English clocks, struck the hours only while continental clocks struck the half hours as well, but the English practice was dropped by some American makers in the middle of the nineteenth century. They fitted an extra tail to the hammer, which the going train raised and then allowed to fall at the half hour, thus indicating the halves without releasing the striking chain.

From this brief historical survey of the American clock industry we see that there were four main types of clocks: tower, tall, wall and shelf clocks.

Tower Clocks

As it has been indicated, the new settlements each expected to have its own clock and presumably the erection of a tower or steeple or public clock was regarded as a sign of respectability, solidarity and permanence. There is an interesting study awaiting someone who can delve into the history of these great clocks and the men who made them. In the Massachusetts town records the following entry appears.

The 9th mo. 1690.

At a Generall towne meeting upon warning, it was agreed that the Bells Capt Crumwell gave the Towne should be by the Selectmen disposed of to the Best Advantage and the produce laid out for one Bell for a Clocke.

What bells? From whence had they come? Did their disposal pay for that much coveted bell in the clock tower? We do not know, but it would be fascinating to find out.

From the *Old Clock Book* by Mrs N. Hudson Moore, we learn that in the year 1704 a tower clock was set up in the steeple of the New Meeting House in Ipswich, Massachusetts.

The reports of these tower or turret clocks are far too fragmentary. There were such clocks in Newburyport as early as 1734 and Benjamin Franklin himself was persuaded to examine one of them after it had been struck by lightning in 1754. In Guildford, Connecticut, Ebenezer Parmalee made a wooden movement for the tower clock in 1726. We assume that Baltimore, Philadelphia, Charleston and almost certainly New York possessed these tower clocks, and among the makers were probably Eli and Samuel Terry, Simon Willard and Seth Thomas. Tower clocks needed maintaining and repairing from time to time and in 1842 Edward Howard founded his own business for this purpose, a business which was to thrive for well over one hundred years.

Tall Clocks

Many elaborate and ornate cases were made in a wide variety of woods for these tall clocks (which were like the English Grandfather Clocks), and they served two purposes. They hid the pendulum and weights, and they kept out the dust. Cases ranged from five to nine feet in height. Records suggest that the tall clocks were popular with all clockmakers, but especially in Pennsylvania where there were at least fourteen of such clocks, and where they continued to be made until the middle of the nineteenth century. Dials of metal, in the earlier models, were about ten and a half inches square. Occasionally extra movements were added to show the phases of the moon and stars, and perhaps a calendar. On the earlier cases there was little ornamentation and this may have been the result of Quaker influence. Later dials were enlarged and sometimes musical devices were added. Some excellent craftsmen, among them the Baileys, Wilders and Benjamin Morrill, produced a smaller edition of the tall clock. Known as the Dwarf, Miniature or Small Tall Clock (and even nicknamed the grandmother clock) they stood only about four feet high. They were obviously not very popular as only a few seem to have been made.

Wall Clocks

The earliest of the wall clocks was the Wag-on-Wall and although many shapes and styles were evolved these clocks were essentially tall clocks without cases. The evolution of the wall clock constitutes an interesting piece of American horological history and the clocks took on many shapes. In the early nineteenth century Simon Willard of Grafton, Massachusetts, produced a number of fine thirty-hour brass movement wall clocks. Clockmakers experimented with case shapes and movements. The Coffin clock was followed by the Banjo, a unique clock shape and movement, patented by Willard in 1862. The Banjo clock is still seen about for it is pleasant in appearance and an accurate timekeeper. Unfortunately the fine case proportions which distinguished this clock have vanished over the years. It is interesting to note that most early Banjo clocks were timepieces, recording time only, and did not strike the hours or contain any alarm mechanism. Later models of the clock were so constructed that they could take two weights, which struck the hours on the casing. Some people may dislike the lack of symmetry about the dial of the Banjo Clock. The winding hole was opposite figure two, thus giving the weight the maximum amount of fall. This unsymmetrical placing of the winding holes may well have been copied from both English and French designs. Some English clockmakers of the seventeenth century had started this practice and it was apparently common among French clocks in the following century. The Presentation Banjo was a Banjo clock set in an ornate gilded and bracketed case. A Banjo-type eight-day brass movement for clock or timepiece was fitted into a design known as the Lyre, and a number of these were successful pieces. Among their makers during the nineteenth century were Aaron Willard Jnr, John Sawin, Dyer (of Sawin & Dyer, Boston, 1820) and Lemuel Curtis. The Girondele, often referred to as America's most beautiful clock, was another Lemuel Curtis design though he probably made no more than about twenty-five.

It was Connecticut again from whence came the major styles of all clocks – the Wag-on-Wall, the Lyre and the Wall

Regulator. The last were used considerably in schools and offices and were fine examples of weight driven regulators. Some are still in existence. About this time, a number of clocks were made in America (though few survive) which were similar to the clocks once seen in practically every government department, office and school in Great Britain. In the early 1800s, under the guidance of Benjamin Morrill, J. Chadwick of Boscowen, USA, and J. Collin, New Hampshire produced its own style of wall clocks. This was an eight-day brass movement affair and was usually a timepiece. One peculiarity about it was the door, which ran the full length of the clock and was divided to show the dial at the top and a mirror below.

Shelf Clocks

As the name shows, these were clocks which were intended to stand on a shelf, mantel or table. They appeared in a wide range of styles, from the delightfully shaped lantern clock of the seventeenth century to the twentieth century alarm clock. Their popularity undoubtedly arose because they could be produced more cheaply and smaller than the tall clock. Not many of the English type bracket clocks were made, the early Americans seemingly preferring the shelf clock known as the Massachusetts Shelf or Half Clock. Simon Willard was one of the makers of this style of clock. The brass movements were both thirty-hour and eight-day and the clocks were simply timepieces or alarm and strike; and excellent timekeepers they were. Some models had kidney dials and others dish dials where the metal dial was convex in shape. The Connecticut clockmakers produced the largest number and greatest variety of shelf clocks. Eli Terry made his first thirty-hour all wood shelf clock about 1808, went on to the Box Case Clock and then the Pillar and Scroll.

This Pillar and Scroll clock was superseded about 1837 by a clock from Chauncey Jerome of Connecticut. It was called the Bronze-looking Glass Clock and was undoubtedly inferior to look at, to say the least. In his autobiography Jerome tells us something about this clock.

Sharp Gothic design by Terry and Andrews, USA, early 19th century

In 1828 I invented the Bronze-looking Glass Clock, which soon revolutionised the whole business. It could be made for a dollar less, and sold for two dollars more, than the Patent Case (the Pillar and Scroll).

Forestville Acorn clock, early 19th century

Jerome's clock had a thirty-hour wooden movement with a taller case to fit the larger pendulum. Inferior in design yet popular, many thousands were made and sold. Had people's tastes slipped, or were they attracted by lower prices for an apparently better clock?

It was about this time that the OG clock appeared, that largest-selling Connecticut-made shelf clock which was still made until 1914. Up to 1837 the movement was of wood but it was then replaced by brass. The clock was made in six or more different sizes and could be bought weight and spring driven with a thirty-hour or eight-day movement. An interestingly shaped clock was the Steeple clock designed by Elias Ingraham. A further unusual type was the Beehive clock.

In New Hampshire some shelf-type clocks with brass movements incorporated what was called the Rat Trap striking mechanism. The locking plate was removed and a series of holes were then drilled in the rim of the great wheel. A wire tapped on the surface of the wheel for each stroke sounded and when it coincided with a hole the train was locked.

AMERICAN WATCHES

We do not know a great deal about the history of the American watch industry and this is a pity, for there is probably an interesting story hidden away up and down the States.

Between 1762 and 1842 Luther Goddard of Shrewsbury, Massachusetts, is believed to have made a number of watches, but his business appears to have been curtailed because the Jefferson embargo prevented the importation of many essential parts. Even when that embargo was lifted Goddard was beaten, for imported watches could sell more cheaply than his. So Goddard cut his losses, gave up watchmaking and returned to preaching the gospel. In Norwich, Connecticut, Thomas Harland apparently had a small factory which turned out two watches and forty clocks a year. This suggests that watchmaking had not really taken on at that stage. Furthermore, we do not know whether Harland's clocks and watches were made from European materials which he imported, though American names were carried on the dials and movements of the final products.

In 1837 the Pitkin brothers of Hartford, Connecticut,

assisted by N. P. Stratton, invented and made their own watch-making equipment in an attempt to increase the manufacture of American watches. Between 800 and 1,000 of the Pitkin watches were made and were described as the quarter plate, slow train size 16. However, competition from abroad was such that within four years the firm was forced out of business. The Waltham Watch Company of Roxbury, Boston (Aaron L. Dennison, 1812–95, and Edward Howard, 1813–1904) was far more successful and survived until 1950. The Elgin National Watch Company of Illinois was another successful American company. It was established in 1864, and in its early days made jewelled movement watches, size 18, full plate, quick train and straightline escapement. It was this company which produced the first stem wind watch and did away with the separate keys formerly required. Watches with keys are now collectors' pieces. Nearly one hundred years later the Elgin Company was still making excellent jewelled movement watches.

Fine American-made watches were on the market by the late 1870s and the industry was at least well established. But the watches were still rather expensive and there was great need for a much cheaper watch. Then in 1880 D. A. Buck of the New Haven Co, Connecticut, after some considerable difficulties announced his Waterbury watch. The watchcase itself was a barrel which held round the modified tourbillon movement something like seven to nine feet of coiled spring. A simplified escapement was incorporated in the watch, and the dial, printed on paper, was covered with celluloid and fastened to the plate. The movement carried the minute hand and made one revolution an hour. It is not surprising that, with only fifty-eight parts and selling at three dollars a time, over 500,000 Waterbury watches were produced. Such a watch, however, was bound to have its problems. Consider, for instance, that enormous length of coiled spring. It took 140 half turns of the stem to wind the watch fully. Stories grew up round this one fact. 'Here, wind my Waterbury for a while. When you are tired I'll finish winding.' The deathknell of this cheap and reliable watch was sounded when certain business concerns began to give a Waterbury watch away free with every suit of clothes sold. Prestige, of course, was bound to fail under such conditions as these.

It was nearly twelve years before another American watch company attempted to market a cheap watch. That company, R. H. Ingersoll of Boston, still a famous worldwide name in horology, had an excellent slogan: 'the watch that made the dollar famous.' Also in 1892 the Hamilton Watch Company was established, and today is one of the greatest manufacturers of jewelled watches in the States.

JAPANESE HOROLOGY

Japanese time measuring originated in China. Before the middle of the nineteenth century Japanese clocks were generally limited to the temples, the rich people and the influential families. This scarcity of time measuring devices may have been due also to the flimsy methods employed in house construction in those early days. Movable partitions, instead of solid interior walls, made the hanging of heavy weight clocks virtually impossible.

The method of time measurement was complicated, and was based upon the natural day, but it began at dusk. The two periods, dusk to dawn and dawn to dusk, were divided into six equal divisions but as the seasons changed the length of the hours differed. Each division, as well as the number attached to it (corresponding to the number of strokes on the bell), was also associated with one of the Chinese signs of the zodiac. In Japan the hours had always been counted backwards, beginning at nine. This number indicated both midnight and midday, and as there were only six periods or hours to be counted, hourly progression was shown as 9...8...7...6...5...4. At the half hours following odd hours a single stroke was sounded, and at the alternate half hours, two strokes. The countwheel of a Japanese striking clock was therefore normally cut to give successively 9...1...8...2...7...1 etc strokes.

In all probability the first western clocks to appear in

Japan were gifts to the Emperor from Dutch trading captains or Portuguese missionaries. The country had been discovered by some Portuguese navigators somewhere about the middle of the sixteenth century, and from the beginning of the next century the Dutch traders made regular visits. Sometime early in the seventeenth century the Japanese began making their own clocks. With their innate imitative ability they were soon able to produce western style clocks but with the dials showing Japanese figures and Chinese symbols for the signs of the zodiac. These clocks measured approximately seventeen inches to the top of the bell and the cases were highly decorated. The Japanese clockmakers were very successful in their attempts to marry the western style clock to their own methods of time-keeping and they produced lantern and bracket clocks. The former, driven by weights, were supported on a stand or could be suspended by a hook from some reasonably permanent support. The latter were spring driven, compact and portable, and they could be stood on any piece of furniture.

However, a clock was built which did not conform to any European type. It was enclosed in a long narrow ornamented case (which held the weights) and could be hung from a permanent wall or pillar in the house. This type of clock, known as a pillar clock, was a good piece of mechanism and many fine examples still exist.

The earlier Japanese clocks were based upon the medieval weight-driven clocks or the English lantern clock. Such clocks possessed one foliot for day and another for night and the change from one to another (at dawn and dusk) was automatic. It was the task of a qualified clockmaker to visit clock owners in order to adjust the foliot weights and to give the clocks a different rate to compensate for the changing length of the hours as the seasons advanced. This was shown by using the falling weight as a pointer to indicate the passing hours. The hours were mounted on a separate plaque and the distance between them could be adjusted in accordance with their length in any given season. This meant that the clock could run at a uniform speed, which in effect meant that there was no need for two sets of escapements or balances. Such a simplified movement resulted in a smaller and more delicate mechanism being produced.

Japanese clockmakers used the balance spring and the small bob pendulum and they made considerable use of the verge escapement. However, for some reason, Western improvements in horology were not taken up and Japanese clockmaking fell behind that of other countries. At the end of the revolution in 1866, which resulted in the restoration of the Mikado and the downfall of the Takugawa dynasty of the Shoguns, Japan opened her doors to Western influences. Within seven years the old complicated method of time keeping had been swept awav and the Western calendar had been accepted. Then clocks of a modern type began to be made in Japan. Probably the first ones were individually made as they had been in America and Europe in the early days, and were consequently still expensive.

The greatest success story in the sphere of Japanese horology is probably that of Seiko. When the Seiko Watch-K. Hattori

Auto-checker (electronic computer for checking accuracy) of Seiko-Hattori of Japan. This latest equipment tests ten watches at a time for accuracy

and Co Ltd began marketing watches just under 100 years ago, the company decided to adopt the policy of providing precise, reliable timepieces. The word *seiko* was aptly chosen for it means 'precision'. By adapting and refining the assembly line method used in other countries and other industries, Seiko has become the world's largest manufacturer of high precision jewelled-lever watches, and they are now being sold all over the world.

During 1969 more than twelve million watches and six million clocks were produced. Something like half the watches and a large quantity of the clocks were exported to over eighty countries. The excellent after-sales service helps to maintain increasing export orders. Seiko's three major manufacturing centres, eighteen assembly plants, and research and development departments are backed by a highly competent engineering, technical and electronics staff. They have together proved that although the days of the individual craftsmen have gone forever, it is still possible to produce precise and sophisticated equipment.

Automatic recessing machine of Seiko-Hattori. Each tiny jewel is recessed automatically by this machine, one of the newest pieces of precision equipment for the controlled production of miniature parts

This accurate photofinish equipment, which records time to 1/100th of a second, is useful in showing the winners in closely contested sports events

All critical parts of watches are machined to a tolerance of one twenty-five thousandth of an inch, a tolerance once unheard of and certainly not possible by hand. Such accuracy was first taken into the sports arena when Seiko-K. Hattori produced a wide range of special new electric and electronic sports timers and systems for the 1964 Olympic Games, which were held in Tokyo. Since that highly successful meeting, many international and intercontinental gatherings, covering practically every type of event, have used the Seiko official timing equipment.

America and Japan have been the only countries outside Britain and Europe to establish their own clockmaking industries and probably America influenced and encouraged the Japanese to enter the factory-made watch and clock business. Both countries have been highly successful.

NAMING TIME

Man likes to name things and even in early days it was realised that reference to past years was only possible if those years were given labels.

The Egyptians used no unit of time longer than a year, and so there was no dating by eras as in modern times. In very early days each year was named after some important event. Later under the fifth and sixth dynasties (about 2494–2345 BC and about 2345–2181 BC) the biennial cattle census was used for time reckoning and the years of a king's reign were numbered alternately as 'the year of the first (second, third, fourth etc) census'. Much later the Egyptians numbered the years of the reign in a straightforward manner (1, 2, 3, 4 etc).

In Assyria from about 890 to the mid-seventh century BC each year was named after a person who held the office of limmu or eponym. An eponym is one who gives his name to a place or institution, and lists of such people were drawn up, the king usually being limmu for one year. Some of these lists have survived. Of the early systems the Assyrian was perhaps the most reliable.

We often say that some event or other took place in a certain era. The first mention of the era to indicate a given period of time was found on an Egyptian monument of Rameses II and was dated in the year 400 of an era which began about 1720 BC. This was rather local and the first era widely known is the Selucid Era; the date it began coincides with Selucid's capture of Babylonia and the foundation of the Selucid Empire.

The Greeks reckoned from the Olympiad of 776 BC and the Romans sometimes used Varros's estimated date of 753 BC when that antiquarian believed that the city of Rome was founded.

However, there is an interesting era we probably give little thought to and it is called the Platonic or Great Year. It is arrived at as follows. The non-spherical earth as it spins is subjected to the gravitational pull of both the moon and the sun. This causes a slight twist to the earth's movement about its axis. The result of this is a slow yet constant change in the axial direction. This creates what is known as the 'precession of the equinoxes'. A complete revolution takes almost 26,000 years and during this long period there is a reversal of winters and summers of both hemispheres.

This book, as its sub-title implies, is merely an introduction to a very wide subject. For those who would like to study it further, a bibliography is included to suggest wider reading.

There is still plenty of scope for research in time measurement, and it can be projected in two opposite directions. To people who are mathematically minded, or have some scientific knowledge of the latest devices, the interest may lie in future developments, particularly of time in space. Others may be content with further enquiry about the past, with the idea of helping to fill in some of the gaps in the record. For example, there is still much to find out about the history of the American clock and watch industry.

Whichever direction, forward-looking or retrospective, we decide to take, we shall find that it leads us to new areas of surprise and interest, and offers further fascinating instances of the ingenuity of man.

Some Interesting Dates in the History of Timekeeping

135 BC	Clepsydra with toothed wheels made by Ctesibius in Alexandria.
AD 850	Pacificus, archdeacon of Verona, is believed to have invented the escapement and the use of weights for motive power.
885	Candles used as clocks. Introduced by Alfred the Great of England.
1326	Oldest clock believed to have been invented by Richard of Wallingford.
1370	Henry de Vic made a clock for Charles V.
1380	Domestic clocks first made in Italy.
1386	Earliest English clock installed at Salisbury Cathedral, England.
1500	Mainspring invented by Peter Henlein of Nürnberg.
1510	Pocket watch first made by Peter Henlein.
1525	Fusee invented by Jacob the Czech.
c 1530	Screws for metalwork came into use for the first time.
1581	Pendulum properties discovered by Galileo the Italian astronomer.
1656	Pendulum clock first designed by Christiaan Huygens the Dutchman.
1658	Pendulum clock first made in England by Fromanteel, who was of Dutch descent.
1660	Balance, or hair, spring invented by the Englishman Dr Robert Hooke.

c 1660 Recoil escapement invented by Dr Robert Hooke.

1671 Pendulum suspension spring introduced by William Clement of Faversham, England.

1673 Mathematics of the pendulum published by Huygens in *Horologium Osciliatorium.*

1680 Seconds' hand made its first regular appearance.

1694 Jewellery in watches first carried out by Nicholas Facio, a Swiss geometrician.

1695 Cylinder escapement invented by Thomas Tompion of London.

1714 British Parliament offered £20,000 prize to anyone who could determine longitude at sea.

1715 Dead beat escapement invented by George Graham, an eminent English clockmaker.

1759 Lever escapement invented by Thomas Mudge of London.

1764 John Harrison awarded part of the £20,000 offered by the British Parliament in 1714.

1765 Compensation balance first made by Pierre Le Roy, Paris.

1780 Marine chronometer escapement perfected by a Londoner, Thomas Earnshaw.

1780 A. P. Breguet of Paris appears as a great horologist.

1780 Automatic winding clock invented by the English clockmaker Louis Recordon.

1840 Electric clocks first made by Alexander Bain of Scotland.

1858 British Horological Institute founded.

1880 Greenwich Mean Time became the Standard Time for the whole of the United Kingdom.

1884 Meridian at Greenwich, as the zero from which longitude is measured, adopted by international agreement.

1895 Dr Ferranti of Liverpool, England, introduced a method of timekeeping from the electric mains.

c 1900 Wristwatches appearing in quantity.

1904 Elinvar metal – compensating metal for balance springs and pendulum rods invented by Dr Charles Guillaume of France.

1912 Summer Time proposed by William Willett of Surrey, England, in his pamphlet *The Waste of Daylight.*

1916 Summer Time introduced in Britain.

1916 Mains plug-in clocks introduced by H. E. Warren of America.

1921 Shortt Free Pendulum invented by William H. Shortt.

1924 Time signals of the British Broadcasting Company (the six pips) first broadcast.

1925 Summer Time Act passed by the British Parliament, although Summer Time was introduced during World War I.

1927 The electricity supply using the 'Grid' system controlled electric clocks.

1929 Quartz crystal clock developed by Dr Warren A. Marrison, USA.

1936 Introduction of TIM by the British Post Office.

1950 Electric transistorised time control system invented by Max Hetzel, a Swiss electronics engineer.

1955 Caesium atomic clock invented by Dr Essen, National Physical Laboratory.

1957 Electric wristwatches first introduced by the Hamilton Watch Company, USA.

1959 The Accutron 'tuning fork' wristwatch, invented by Max Hetzel and made by the Bulova Co, USA.

1970 Atomic wristwatch incorporating the radio isotope of Promethium 147 developed by McDonnell Douglas, an American astronautics company, to the specifications of Hattori of Japan.

Bibliography

Baillie, G. H., *Watchmakers and Clockmakers of the World.* NAG Press, 1929.

Baillie, G. H., *Clocks and Watches.* NAG Press, 2nd edition, 1947.

Benson, J. W., *Time and Time-tellers.* Benson, 1902.

Britten, F. J. *Old Clocks and Watches and their Makers.* Batsford, 1933. Revised Spon, 1956.

de Carle, D., *British Time.* Lockwood, 1947.

de Carle, D., *Practical Watch Repairing.* NAG Press, 1952.

Edey, Winthrop., *French Clocks.* Studio Vista, 1967.

Essen, L., *Standards of Time and Frequency.* Reprint from *Research,* Vol. 10, June 1957.

Essen, L., *Atomic Clocks.* Reprint from *Research,* Vol. 15, June 1962.

Gordon, G. F. C. *Clockmaking Past and Present.* Technical Press, 1946.

Gould, R. T., *The Marine Chronometer.* Potter, 1923.

Hattori, K., *The Seiko Story.* Hattori, Tokyo, 1970.

Hood, P., *How Time is Measured.* OUP, 1955.

Palmer, Brooks., *A Treasury of American Clocks.* Collier-McMillan, 1967.

Palmer, Brooks., *A Book of American Clocks.* Collier-McMillan, 1950.

Priestley, J. B., *Man and Time.* Aldus Books, 1964.

Robertson, J. D., *The Evolution of Clockwork* (with special section on the clocks of Japan). Cassell, 1931.

Symonds, R. W., *A Book of English Clocks.* Penguin, 1947.

Toulmin, S. and Goodfield, J., *The Discovery of Time.* Pelican, 1967.

Tremayne, A., *Everybody's Watches – Everybody's Clocks.* NAG Press, 1949.

Tyler, E. J., *European Clocks*. Ward, Lock & Co Ltd, 1968.
Ward, F. A. B., *Time Measurement (Part I) – Historical Survey*. HMSO, 1949.
Ward, F. A. B., *Time Measurement (Part I) – Historical Catalogue*. HMSO, 1950.
Watkins, H., *Time Counts*. Spearman, 1954.
Way, R. B. and Green, N. D. *Time and its Reckoning*.

Acknowledgements

I should like to express my gratitude to the following for the advice and help they have so generously given me during the preparation of this work:

Dr C. F. Beeson, editor of the *Antiquarian Horological Society Journal*.
E. Dent & Co Ltd for information about the history of Big Ben.
Dr L. Essen of the National Physical Laboratory, Teddington, for valuable assistance in connection with the quartz and atomic clocks.
Mr R. W. Symonds, FRIBA, for making valuable comments upon the manuscript in its early stages.
Mr Charles L. Granquist, Smithsonian Institution, Washington D.C., USA.
Mr Humphry M. Smith, BSc, FInstP, MIEE, FRAS, Officer-in-Charge, Time Department, Royal Greenwich Observatory.
Mr A. W. Marshall, of Smiths English Clocks Ltd.
Hattori, Tokyo.

K. F. Welch
Dorchester-on-Thames 1971

Index

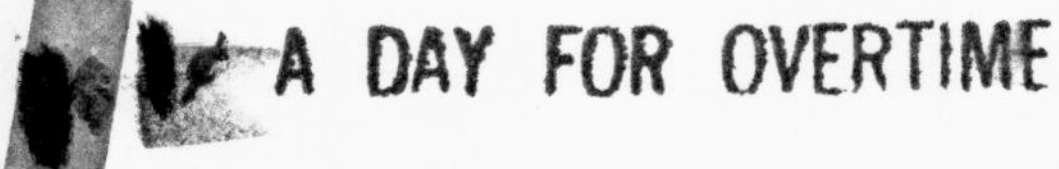
A DAY FOR OVERTIME